Critical Views of Logic

This book examines positions that challenge the Fregean "logic-first" view. It raises critical questions about logic by examining various ways in which logic may be entangled with mathematics and metaphysics.

Is logic topic-neutral and general? Can we take the application of logic for granted? This book suggests that we should not be dogmatic about logic but ask similar critical questions about logic as those Kant raised about metaphysics and mathematics. It challenges the Fregean logic-first view according to which logic is fundamental and hence independent of any extra-logical considerations. Whereas Quine assimilated logic and mathematics to the theoretical parts of empirical science, the present volume explores views that stop short of his thoroughgoing holism but instead take logic to be answerable to or entangled with some particular disciplines. The contributions provide views that assign primacy to mathematical reasons, Kantian metaphysical grounds, Husserlian transcendental phenomenological reflection, or normative considerations about how terms ought to be defined in various fields of empirical science or mathematics. Space is thereby carved out between the Fregean position on the one hand and Quinean holism on the other.

Critical Views of Logic will be a key resource for academics, researchers, and advanced students of philosophy, linguistics, mathematics, and computer science, as well as those engaged in various fields of empirical science. The chapters in this book, except for Chapter 4, were originally published in the journal *Inquiry*.

Mirja Hartimo (University of Helsinki, Finland) works on the phenomenological perspective on philosophy of mathematics and logic. Her recent publications include *Husserl and Mathematics* (2021).

Frode Kjosavik (Norwegian University of Life Sciences, Norway) works on Kant, Husserl, metaphysics, philosophy of mathematics, and philosophy of science. He is the editor of *Metametaphysics and the Sciences: Historical and Philosophical Perspectives* (with Camilla Serck-Hanssen, 2020).

Øystein Linnebo (University of Oslo, Norway) works on metaphysics and the philosophy of logic and mathematics. His recent publications include *Thin Objects: An Abstractionist Account* (2018) and *The Many and the One: A Philosophical Study of Plural Logic* (with Salvatore Florio, 2021).

Critical Views of Logic

Edited by
Mirja Hartimo, Frode Kjosavik and
Øystein Linnebo

LONDON AND NEW YORK

First published 2024
by Routledge
4 Park Square, Milton Park, Abingdon, Oxon, OX14 4RN

and by Routledge
605 Third Avenue, New York, NY 10158

Routledge is an imprint of the Taylor & Francis Group, an informa business

Introduction © 2024 Mirja Hartimo, Frode Kjosavik and Øystein Linnebo
Chapters 1–3, 5–9 © 2024 Taylor & Francis

Chapter 4 © 2020 Salvatore Florio and Øystein Linnebo. First published in the journal *Philosophia Mathematica*, 28.2 (2020). Reprinted with permission of Oxford University Press.

All rights reserved. No part of this book may be reprinted or reproduced or utilised in any form or by any electronic, mechanical, or other means, now known or hereafter invented, including photocopying and recording, or in any information storage or retrieval system, without permission in writing from the publishers.

Trademark notice: Product or corporate names may be trademarks or registered trademarks, and are used only for identification and explanation without intent to infringe.

British Library Cataloguing-in-Publication Data
A catalogue record for this book is available from the British Library

ISBN13: 978-1-032-57351-9 (hbk)
ISBN13: 978-1-032-57354-0 (pbk)
ISBN13: 978-1-003-43899-1 (ebk)

DOI: 10.4324/9781003438991

Typeset in Myriad Pro
by codeMantra

Publisher's Note
The publisher accepts responsibility for any inconsistencies that may have arisen during the conversion of this book from journal articles to book chapters, namely the inclusion of journal terminology.

Disclaimer
Every effort has been made to contact copyright holders for their permission to reprint material in this book. The publishers would be grateful to hear from any copyright holder who is not here acknowledged and will undertake to rectify any errors or omissions in future editions of this book.

Contents

   Citation Information   vii
   Notes on Contributors   ix

   Introduction to "Critical views of logic"   1
   Mirja Hartimo, Frode Kjosavik and Øystein Linnebo

1   Infinity and a critical view of logic   9
   Charles Parsons

2   Dummett's objection to the ontological route to intuitionistic logic: a rejoinder   28
   Mark van Atten

3   The entanglement of logic and set theory, constructively   46
   Laura Crosilla

4   Critical plural logic   68
   Salvatore Florio and Øystein Linnebo

5   Kant on the possibilities of mathematics and the scope and limits of logic   101
   Frode Kjosavik

6   The infinite, the indefinite and the critical turn: Kant via Kripke models   125
   Carl Posy

7   Husserl on Kant and the critical view of logic   156
   Mirja Hartimo

8   Logical pluralism and normativity   174
   Teresa Kouri Kissel and Stewart Shapiro

9 Disagreement about logic   194
   Ole Thomassen Hjortland

   Index   217

Citation Information

The following chapters, except for Chapter 4, were originally published in various volumes and issues of the journal *Inquiry*. Chapter 4 was originally published in the journal *Philosophia Mathematica*, volume 28, issue 2 (2020). When citing this material, please use the original page numbering for each article, as follows:

Chapter 1
Infinity and a Critical View of Logic
Charles Parsons
Inquiry, volume 58, issue 1 (2015) pp. 1–19

Chapter 2
Dummett's objection to the ontological route to intuitionistic logic: a rejoinder
Mark van Atten
Inquiry, volume 65, issue 6 (2022) pp. 725–742

Chapter 3
The entanglement of logic and set theory, constructively,
Laura Crosilla
Inquiry, volume 65, issue 6 (2022) pp. 638–659

Chapter 4
Critical Plural Logic
Salvatore Florio and Øystein Linnebo
Philosophia Mathematica, volume 28, issue 2 (2020) pp. 172–203

Chapter 5
Kant on the possibilities of mathematics and the scope and limits of logic
Frode Kjosavik
Inquiry, volume 65, issue 6 (2022) pp. 683–706

Chapter 6
The infinite, the indefinite and the critical turn: Kant via Kripke models
Carl Posy
Inquiry, volume 65, issue 6 (2022) pp. 743–773

Chapter 7
Husserl on Kant and the critical view of logic
Mirja Hartimo
Inquiry, volume 65, issue 6 (2022) pp. 707–724

Chapter 8
Logical pluralism and normativity
Teresa Kouri Kissel and Stewart Shapiro
Inquiry, volume 63, issue 3–4 (2020) pp. 389–410

Chapter 9
Disagreement about logic
Ole Thomassen Hjortland
Inquiry, volume 65, issue 6 (2022) pp. 660–682

For any permission-related enquiries please visit:
http://www.tandfonline.com/page/help/permissions

Notes on Contributors

Laura Crosilla, Department of Philosophy, Classics, History of Art and Ideas, University of Oslo, Norway.

Salvatore Florio, Department of Philosophy, Classics, History of Art and Ideas, University of Oslo, Norway.

Mirja Hartimo, Department of Philosophy, History and Art Studies, University of Helsinki, Finland.

Ole Thomassen Hjortland, Department of Philosophy, University of Bergen, Norway.

Frode Kjosavik, School of Economics and Business, Norwegian University of Life Sciences (NMBU), Norway.

Teresa Kouri Kissel, Department of Philosophy and Religious Studies, Old Dominion University, Norfolk, USA.

Øystein Linnebo, Department of Philosophy, Classics, History of Art and Ideas, University of Oslo, Norway.

Charles Parsons, Department of Philosophy, Harvard University, Cambridge, USA.

Carl Posy, Department of Philosophy, The Hebrew University of Jerusalem, Israel.

Stewart Shapiro, Department of Philosophy, Ohio State University, Columbus, USA.

Mark van Atten, Archives Husserl (CNRS/ENS), Paris, France.

Introduction to "Critical views of logic"

Mirja Hartimo, Frode Kjosavik and Øystein Linnebo

This volume explores what we call "critical views of logic." Following Frege, logic is often regarded as epistemologically and methodologically fundamental. All disciplines—including mathematics—are answerable to logic rather than vice versa. Critical views of logic disagree with this "logic-first" view. The logical principles that govern some subject matter may depend on the metaphysics of this subject matter or on the semantics of our discourse about it.

Challenging the logic-first view

According to Frege, logic codifies "the basic laws" of all rational thought, and the laws of logic must therefore be presupposed by all other sciences. What, then, might justify a law of logic? We could of course consult logic itself:

> As to the question, why and with what right we acknowledge a logical law to be true, logic can respond only by reducing it to other logical laws. Where this is not possible, it can give no answer. (Frege 1893 [2013], p. xvii)

But this will not take us very far, for logic too will need some fundamental laws. Frege therefore continues by asking whether there are extralogical considerations to which we can appeal:

> Stepping outside logic, one can say: our nature and external circumstances force us to judge, and when we judge we cannot discard this law—of identity, for example—but have to acknowledge it if we do not want to lead our thinking into confusion and in the end abandon judgement altogether. I neither want to dispute nor to endorse this opinion, but merely note that what we have here is not a logical conclusion. What is offered here is not a ground of being true but of our taking to be true. (*ibid*.)

Thus, there is no help to be had from extralogical considerations either. Whereas in the *Grundlagen* (Frege 1884 [1953]) Frege seemed attracted to the idea that logic is constitutive of our thinking or judging, he is now unwilling to endorse this as a reason for *the truth* of the laws of logic, seeing it only as a reason for *our taking* the laws of logic to be true. It will also not help to look to other sciences:

> I take it to be a sure sign of error should logic have to rely on metaphysics and psychology, sciences which themselves require logical principles. (*ibid.*, p. xix)

In the absence of any help from other sciences, the only option is—as Wittgenstein put it two decades later—that "Logic must take care of itself."[1] Let us call this a *logic-first view*.

What we call "critical views of logic" challenges this Fregean logic-first view. As Heyting observes, although the idea is older, Brouwer was the first thinker to properly develop it:

> The idea that for the description of some kinds of objects another logic may be more adequate than the customary one has sometimes been discussed. But it was Brouwer who first discovered an object which actually requires a different form of logic, namely the mental mathematical construction. (Heyting 1956, 1)

Although intuitionists defend a critical view of logic, they are not alone in doing so. Here we follow Charles Parsons (2015), who has served as an important source of inspiration for us. Parsons characterizes a critical view of logic as one that does not take logic for granted but considers it to be open to substantive revisions. As early proponents, he mentions Brouwer, Weyl, and Hilbert. Their critique is partly based on reflections on specific facts about the domain (such as its infinite size), which lead to a reconsideration of certain logical laws. For Parsons, Bernays represents the view in a form that is especially close to his heart. The aim of this volume is to explicate and elaborate on various critical views of logic.

What kind of criticism of logic is possible? As we observed, a minimal form of criticism is possible even according to the Fregean view, namely *logic-internal criticism*, in which logic takes care of itself, as Wittgenstein put it. In Kant as well, there are arguably similar moves. Kant was in many ways concerned with giving a correct account of logic. In that sense, he took a critical view of Aristotle's logic and of contemporary logic. For example, in the pre-critical period, he wrote the "False Subtlety of the Four Syllogistic Figures" (1762). In his critical period, he considered, e.g., whether there ought to be a distinction between singular and universal judgments within general logic.

What we are after here, however, are views that are critical of logic in a more radical way than this logic-internal criticism. It is the sense in which Kant is "critical" rather than "dogmatic" about traditional metaphysics. Parsons too emphasizes that he intends to use the term "critical" in a sense

---

[1] This is a recurring theme in his *Notebooks 1914–1916* (Wittgenstein 1979); see especially the first three entries. The claim is reiterated in the *Tractatus* (Wittgenstein 1921 (1922), 5.473).

akin to the Kantian one, only applied to logic rather than metaphysics. He does not ascribe a critical view of *logic* to Kant.

The critical views of logic are thus not dogmatic about logic. Hence, they are not logic-first. What, then, is logic answerable to or entangled with? There are many possibilities. A radical departure from a logic-first view, which regards logic as "given," is Quine's thoroughgoing holism, which assimilates logic and mathematics to the theoretical parts of empirical science. These disciplines, Quine claims, are not essentially different from theoretical physics: although they go beyond what can be observed by means of our unaided senses, they are justified by their contribution to the prediction and explanation of states of affairs that can be thus observed.

This collection is particularly concerned with less radical departures from Frege, which stop short of Quine's thoroughgoing holism but instead take logic to be answerable to or entangled with some particular disciplines, such as mathematics, semantics, or some part of metaphysics. For example, according to Brouwer, logic is answerable to mathematics, while Dummett takes logic to be answerable to the theory of meaning. The contributions in this volume provide examples of various other views that deviate from Frege's: views that assign primacy to anything from Kantian metaphysical grounds or constructivist mathematical reasons to Husserlian transcendental phenomenological reflection or normative considerations about how terms ought to be defined in various fields of mathematical practice or empirical science. The critical views of logic discussed here carve out some interesting space between the Fregean logic-first view and thoroughgoing holism *à la* Quine (which are also receiving renewed attention, as described in, and evidenced by, Hjortland 2017).

We consider these intermediate views—profiled in this volume—to be the most interesting and promising. On the critical views, the entanglement of logic with other disciplines is local, not global as on Quinean holism. Logic might, for example, be entangled with mathematics. Such a localized entanglement can be deeper—and therefore also more informative and more interesting—than the somewhat bland view that everything belongs to one vast web of beliefs.

Varieties of logic entanglement—chapter outline

The opening chapter is **Charles Parsons**' "Infinity and a Critical View of Logic," which has served as the original source of inspiration for the entire collection. Parsons explores what he calls the "entanglement of logic and mathematics" that comes to the fore in various ways when we apply logic to reasoning involving infinities. After a survey of some earlier attempts at a critical view of logic from Brouwer onward, he illustrates the entanglement he has in mind in the context of second-order logic, where it in his view has some "rather problematic" aspects. He moves on to raise the question of the correctness of first-order logic in application to the higher infinite. He suggests that Brouwer's and others' critique of applying logic to reasoning about the infinite should be taken as a signal, not so much to adopt constructive mathematics or to otherwise "blame" mathematical assumptions but to worry about applying first-order logic to reasoning about infinite totalities, for example, in analysis and in higher set theory. Parsons' critical view of logic crystallizes in his final questions: "How well do we understand logical validity in the context of reasoning about the higher infinite? Do we really understand it better than we understand the higher infinite itself?"

In the subsequent chapters, the entanglement of logic and mathematics is explored from various points of view. In so far as logic is considered to be entangled with mathematics, mathematical facts need to be taken into account when laying down the logical principles for the domain in question. **Mark van Atten**'s and **Laura Crosilla**'s contributions are examples of critical views of logic in this "mathematics-first" sense, but with a more constructivist view of these facts—with their specific ontological justifications or implications.

Van Atten considers Michael Dummett's claim that there is no ontological route to an intuitionistic logic for elementary number theory, only a general meaning-theoretic route, which does not rely on any specific construal of mathematical meanings. Even if it is granted that natural numbers are creations of human thought, Dummett argues, decidable statements about them that are not yet decided can be taken as having a determinate truth-value prior to the actual execution of the decision procedures. Hence, an anti-Platonist ontology leaves room for classical logic for such statements and for the principle of bivalence. Drawing his inspiration from Brouwer, van Atten defends the ontological route to intuitionistic logic from this criticism. According to ontological intuitionism, proof procedures are *generative* rather than *investigative* and thus bring about that by virtue of which a mathematical statement about a natural number is true. Hence, if we do not yet know whether a natural number is prime or composite, a decision procedure for primality either *makes it prime* or *makes it composite*. It remains true that were we to carry out the decision procedure, the natural number would turn out one way or the other, i.e., intuitionistic and classical logic "formally coincide" for decidable statements. However, the interpretations differ, and on generative views of proofs the statement that the natural number is prime does not already have a determinate truth-value.

Whereas van Atten focuses on elementary number theory, Crosilla is concerned with set theories, with their far larger infinite domains. Logical principles cannot be cleanly separated from set theoretic axioms, she shows. This calls into question the alleged topic-neutrality of logic. In particular, the standard versions of the axioms of foundation and choice in ZF can be seen to entail classical logic, which brings out the entanglement between logic and sets. This has implications for how the axioms of a constructive set theory ought to be formulated. She then turns to Poincaré's way of avoiding impredicativity and actual infinity through the definitions of sets. This is seen as a variant of the view that sets are extensions of concepts; it also suggests genetic definitions. Finally, Martin-Löf type theory makes sense of quantification over infinite domains through a shift from classical to intuitionistic logic, along with predicativity. In this theory, there is even more of an entanglement between logic and mathematics than in constructive ZF. Logic and sets are not introduced separately, and propositions are identified with types of their proofs, i.e., with sets—this is the Curry-Howard isomorphism. It turns out that the critique of logic in question makes more precise sense of Poincaré's approach as well.

As Parsons pointed out, realists too can take a critical view of logic. In this vein, **Salvatore Florio** and **Øystein Linnebo** develop and defend a "critical" plural logic that restricts the traditional logic of plurals, for example, by not warranting a universal plurality of all objects whatsoever. The central premise of their argument is a liberal view of permissible mathematical definitions. Suppose we wish to add certain "new" objects to some "old" objects that we have already accepted. Suppose further that the desired "new" objects can be fully characterized in terms of the "old" ones. Then the desired new objects are permissible. For example, consider two pluralities *xx* and *yy* of "old" objects. Then the associated sets {*xx*} and {*yy*} can be fully characterized in terms of the "old" objects *xx* and *yy*. This view of definitions entails that every plurality of objects we accept (which therefore count as "old") defines a set. Consequently, there can be no universal plurality since such a plurality could be used to define a set, whose existence would contradict this plurality's supposed universality. Unlike the traditional logic of plurals, critical plural logic is (provably) compatible with the liberal view of definitions. Which plural logic, then, should we accept—the traditional one or the "critical" alternative? A logic-first approach to the question would be unhelpful and dogmatic. We confront a choice between two package deals: (i) traditional plural logic and a restricted view of permissible mathematical definitions or (ii) critical plural logic and an attractively general theory of such definitions. The choice of a plural logic thus becomes entangled with questions in the philosophy and foundations of mathematics. As Williamson (2018, p. 95) observes, such choices should be decided based on which alternative is "more informative, with more power to unify and explain general patterns." In light of the mentioned entanglement, Florio and Linnebo argue that the critical option should prevail.

The entry point for **Frode Kjosavik** and **Carl Posy** is not an entanglement of logic with mathematics but an entanglement with full-fledged metaphysics or the critique thereof, in particular as this critique can be found in Kant's transcendental philosophy. Both authors concentrate on the first and second antinomies of rational cosmology in the *Critique of Pure Reason* and the considerations of infinite domains contained in these.

Kjosavik points out that, while Kant denied that *reductio ad absurdum* proofs could be used to settle traditional metaphysical issues, he did not take such proofs to be invalid as such. To set the reasoning in the antinomies straight, however, a propositional version of the law of excluded middle has to be supplemented with a predicative version. There are resources for this in Kant's own formal logic, in the form of the distinction between negative and "infinite" judgments. Thus, what is subjected to critique is classical logic's restriction of negation to contradictory opposition between propositions. The law of excluded middle is not criticized as such. Moreover, if mathematics is extended beyond the constraints that Kant imposed on it, analogues of the metaphysical antinomies emerge within mathematics as well, thereby extending the scope of a Kantian critique of classical logic to mathematics. It is argued that Zermelo's own take on the Kantian antinomies in connection with the hierarchy of sets is particularly helpful here, in that it brings out the fruitfulness of relaxing Kant's finitary constructability condition in mathematics.

Posy, on the other hand, argues that a shift from classical to intuitionistic logic is required to make sense of the apparent contradictions. Within mathematical discourse, Kant takes the law of excluded middle to be universally valid. Even if there is no proof or refutation of a proposition, or any recognition of a way of proving or refuting it, the required piece of evidence can in principle always be found since it comes from within ourselves. From this "standpoint of Reason," it can be asserted of any statement that it is either provable or refutable independently of actual evidential growth. Within cosmology—now in the form of rationalist metaphysics transformed into empirical science—this is not so. There is no in-principle rational decidability here. Nor can there ever be conclusive evidence for the cosmological questions raised in the first and second antinomies. Notably, Brouwer, unlike Kant, adopts this "standpoint of the Understanding" even for mathematics, i.e., that of restriction of warranted assertability to what can be recognized at evidential situations. Posy employs intuitionistic Kripke models to illuminate the valid reasoning that ultimately comes out of the two Kantian antinomies. Different models capture Kant's finer distinction between evidential states that progress *ad indefinitum* and those that progress *ad infinitum*, which is also linked to infinity considerations in the metaphysics of Descartes and Leibniz. Furthermore, a separate metalogic is suggested through Posy's construal of the two standpoints. In the metatheory, the Kripke models are to be viewed classically, from the standpoint of Reason.

According to **Mirja Hartimo**, Husserl has an even broader entanglement view of logic than the ones presented so far. His view is not that of entanglement with mathematical facts-cum-ontology, as in van Atten and Crosilla. Nor is primacy accorded to entanglement with metaphysics, or the critique thereof, as in Kjosavik and Posy. Rather, logic is part of a package where both mathematics and metaphysics enter. First, formal logic comprises not only principles for reasoning but, more fundamentally, all of pure mathematics—including Hilbert's axiomatics and the Riemannian theory of manifolds —and thereby "formal ontology." Second, it is normative standards—not facts about a subject matter, such as infinite domains—that determine which logical principles are to be adopted for a particular discourse. Third, what is effectively Husserl's take on traditional metaphysics leads to a critical view that consists in much more than an undogmatic approach to substantive logical issues; rather, the view requires us to seek answers to the "How is x possible"-question for formal logic—a question that never extended that far in the work of Kant. Husserl thereby equates being critical of logic with seeking "transcendental-phenomenological" foundations for a logic that is descriptive of ideal formations. In the end, however, Hartimo proposes a position where considerations of mathematical facts also play an important part, as they do in Parsons.

Teresa Kouri-Kissel's and **Stewart Shapiro**'s contribution adds Carnap's point of view to the range of critical views of logic and explores its ramifications for the normativity of logic. To be sure, for them, Carnap's logical tolerance is an outcome of rational reconstruction of underlying practice of mathematics or science. Since there are legitimate mathematical or scientific theories that involve different logics, the authors argue, the idea of One True Logic that holds for any subject matters should be abandoned. Thus, their contribution develops a critical view of logic which is pluralist or, perhaps we should say, *localist* about logic. The consequent view of normativity of logic is likewise antithetical to Fregean monism. The authors argue that various logics are rational reconstructions of the norms implicit in the practices of classical analysis, intuitionistic analysis, inconsistent mathematics, and so forth. In their view, each of the logics is truth-preserving in its domain, but none of them, or perhaps only the weakest of them, are truth preserving "across the board."

Ole Hjortland, in turn, defends the position of entanglement between logic and science in general, as in Quine's gradualism doctrine, according to which logic is continuous with the empirical sciences and therefore amenable to revisions. All positions that give logic an exceptional status are thereby rejected. This includes not only apriorism and foundationalism but also, somewhat intriguingly, a view often ascribed to Quine himself, namely that any disagreement about logic is symptomatic of a disagreement about meaning, such that there can be no genuine disagreement about logic and hence also no genuine critique thereof. To replace classical logic with a deviant one, say, by rejecting the law of excluded middle, is to change the subject or revise the usage of terms. Hjortland argues, however, that there are normative issues within any science concerning what its central terms *ought* to mean. Furthermore, our choice of concepts and of theories cannot be divorced from each other. Deep disputes over how terms ought to be defined or explicated can lead to alternative logics and thus represent critical views of logic, i.e., substantial rather than merely verbal disagreements.

This also offers a way to reconcile the claim that logical disputes are about meaning with a view according to which logic is continuous with empirical sciences and open to rational revisions.

References

Frege, Gottlob. 1884. *Die Grundlagen der Arithmetik: Eine logisch mathematische Untersuchung über den Begriff der Zahl*. Breslau: W. Koebner. 1953. *The Foundations of Arithmetic*. Transl. by J. L. Austin. 2nd revised edn. Oxford: Blackwell.

Frege, Gottlob. 1893. *Grundgesetze der Arithmetik*. Jena: Pohle. Vol. I. Transl. by P. A. Ebert and M. Rossberg. 2013. *Basic Laws of Arithmetic*. Oxford: Oxford University Press.

Heyting, Arend. 1956. *Intuitionism: An Introduction*. Amsterdam: North-Holland.

Hjortland, Ole. 2017. "Anti-Exceptionalism about Logic," *Philosophical Studies* 174(3): 631–658.

Parsons, Charles. 2015. "Infinity and a Critical View of Logic," *Inquiry* 58(1): 1–19.

Williamson, Timothy. 2018. *Doing Philosophy: From Common Curiosity to Logical Reasoning*. Oxford: Oxford University Press.

Wittgenstein, Ludwig. 1921. *Logisch-philosophische Abhandlung, Annalen der Naturphilosophie* 14, 185–262. 1922. *Tractatus Logico-Philosophicus* [TLP]. Transl. by C. K. Ogden. London: Routledge.

Wittgenstein, Ludwig. 1979. *Notebooks 1914–1916*, 2nd edn. G. H. von Wright and G. E. M. Anscombe (eds.). Oxford: Blackwell.

Infinity and a critical view of logic

CHARLES PARSONS

ABSTRACT   *The paper explores the view that in mathematics, in particular where the infinite is involved, the application of classical logic to statements involving the infinite cannot be taken for granted. L. E. J. Brouwer's well-known rejection of classical logic is sketched, and the views of David Hilbert and especially Hermann Weyl, both of whom used classical logic in their mathematical practice, are explored. We inquire whether arguments for a critical view can be found that are independent of constructivist premises and consider the entanglement of logic and mathematics. This offers a convincing case regarding second-order logic, but for first-order logic, it is not so clear. Still, we ask whether we understand the application of logic to the higher infinite better than we understand the higher infinite itself.*

Some of the classic thinkers on the foundations of mathematics in the early twentieth century maintained that the applicability of the usual logical laws to reasoning about the infinite can be questioned. One of them, L. E. J. Brouwer, famously maintained that some of these laws should be rejected. However, my focus will be on what Brouwer's view of logic had in common with that of contemporaries who, in practice at least, upheld classical mathematics. Because Brouwer is so well known, I will treat his views more briefly and then deal with Hermann Weyl and David Hilbert. They provide a reason for describing the view of logic that concerns me as 'critical', echoing Kant, who contrasted his own position with dogmatism and skepticism. In the last part of the paper, I will consider the issue from a more contemporary point of view.

A view that has been natural to many thinkers is that logic is evident in some autonomous way either because its basic principles are self-evident or because it is so basic that it cannot be a sensible question what the evidence of clearly correct logical inferences is based on. The term 'self-evident' is usually understood in a maximal way, so that if two propositions are both self-evident, they are equally evident. The same would be a natural reading of the view that there is no room for sensible questions about the basis of the evident character of logical inferences. It would then follow that if a fragment of logic is codified as a formal system, none of the axioms or basic inferences would be singled out as more evident than the others, or for that matter as less evident. Or if there are such differences, it is a 'psychological' matter; the 'more evident' principles are easier for our minds to assimilate, at least at a given stage of history.

Both of the views sketched in the last paragraph have been ascribed to Frege. An especially clear version of the self-evidence view is presented in writings of his more rationalist interpreters, such as Tyler Burge.[1] However, Frege himself qualified that position on one point, in admitting that even before Russell's paradox came to light, he thought that his Basic Law V was not as evident as a logical principle should be. However, he was not led by this debacle to question what he called 'fundamental logic', close enough for our purposes to axiomatic second-order logic. I interpret the view that the basic principles of logic are self-evident as implying that their evident character can be assumed in mathematical reasoning, in particular reasoning about the infinite. Apart from epistemological views about logic, it has been common in American philosophy of mathematics to take the soundness of the existing logic, at least classical first-order logic, for granted, in particular in its application in mathematics.

The view of logic that I am most concerned to explore is opposed to the views and attitudes just sketched. Since its pioneer was Brouwer, and it was a key element in Brouwer's intuitionism from early on, it is in a way familiar. But I propose to present it in a way that will make clear that intuitionists were not the only ones to hold it. In particular, we can find it in Hilbert's public lectures of the 1920s, when Brouwer and Hilbert were in the course of becoming bitter rivals. We also find it in Hermann Weyl's writings on the foundations of mathematics of the same period. In sections 2–4, I will discuss their views, giving more attention to Weyl, because he is less well known and because some of his formulations are especially vivid.

These three were probably the primary exponents of the view that concerns me, but it was probably quite widely held in the 1920s and into the 1930s. It is not forgotten today, because the period in which these three mathematicians flourished continues to be studied. But it is not so clear whether the view has any influence on the foundational debates of recent times. The question of its contemporary relevance will be addressed at the end of the paper.

The basic idea of the view is that we cannot take for granted the familiar logical principles and inferences in doing mathematics, in particular when our reasoning involves the infinite, even in the very low-level way in which the infinite enters into reasoning about natural numbers. That, for a given predicate

---

[1] See especially his 'Frege on Knowing the Foundation'.

F, either 'Every number is F' or 'Some number is not F' must be true is an assumption, one should say a substantive mathematical assumption, in which it is possible to contest or to do without. This simple example indicates that the view I am concerned with questions even first-order logic.[2] For Weyl and Hilbert, the understanding of quantifiers is especially central. Although Brouwer's target was not a view of quantifiers, we will see that quantifiers are central to some typical arguments of his.

I. Brouwer

Brouwer was obviously the first to press the idea that basic logic in application to mathematics can be questioned. He did so before first-order logic was clearly singled out, and his procedure was always informal, not dependent on a single formalization. His target was almost always the law of excluded middle and other principles that belong to propositional logic. From our point of view, the examples would be read as involving quantifiers, and he provided an ample supply of them that we would read as first order.

To see this, I will mention a simple one from a relatively early paper. The proposition at issue is that 'the points of the continuum form an ordered point species', which obviously implies that for every such point r, either $r > 0$ or $r = 0$ or $r < 0$. Let k be the least number m, if it exists, such that

> the segment $d_m\,d_{m+1}, \ldots d_{m+9}$ of the decimal expansion of π forms the sequence 0123456789. Further, let $c_n = (-1/2)^k$ if $n \geq k$, otherwise let $c_n = (-1/2)^m$; then the infinite sequence $c_1, c_2, c_3, \ldots$ defines a real number r for which none of the conditions $r = 0$, $r > 0$, or $r < 0$ holds.[3]

In other words, $r = 0$ if there is no sequence 0123456789 in the decimal expansion of π, $= (-1/2)^k$ if k is the least m initiating such a sequence, so that r is positive or negative according as k is even or odd.

In a rather schematic presentation of examples of this kind in his well-known philosophical essay 'Mathematik, Wissenschaft, und Sprache', Brouwer makes clear how examples are generated from what we would call Σ_1^0 statements that we see no way of deciding. The generality is important, because as

---

[2]It might seem from the example that already monadic first-order logic is questioned, but that appearance is misleading, because actual reasoning about natural numbers quickly becomes polyadic. I don't think the question whether monadic first-order logic can escape criticism is very interesting. However, the fact that a monadic formula that is not valid has a finite countermodel does indicate that monadic logic is less entangled with mathematics than full first-order logic.

[3]'Über die Bedeutung des Satzes vom ausgeschlossenen Dritten', 3, notation slightly modified.

Brouwer remarks, a particular unsolved problem like the one in the example just given might be solved.[4]

Brouwer defines a *fleeing property* as a decidable property for which

> ... one cannot calculate a particular number that has the property, nor can one prove the absurdity of the property for all natural numbers. We define the critical number [*Lösungszahl*] λ_f of a fleeing property f as the (hypothetical) smallest natural number that possesses the property; we further define an up number and a down number of *f* as a number that is, respectively, not smaller and smaller than the critical number. It is immediately clear that for an arbitrary fleeing property each natural number can be recognized to be either an up number or a down number and that in the first case the property loses its character as a fleeing property.[5]

A fleeing property is *parity-free* if one does not have a proof either that no number satisfying *f* is even or that no number satisfying *f* is odd.[6] For a parity-free fleeing property, the binary oscillating number (*duale Pendelzahl*) is the real number r determined by the Cauchy sequence $a_1, a_2, a_3,...$, where for an arbitrary down number n of f, a_n is $(-1/2)^n$, and for an arbitrary up number of f, a_n is $(-1/2)^k$, where $k = \lambda_f$. Clearly, $r = 0$ if there is no n satisfying f and is λ_f if there is one. Brouwer says that neither $r = 0$ nor $r \neq 0$ holds, contrary to the law of excluded middle. Evidently, what he means is that we cannot assert that $r = 0$ or $r \neq 0$. Call a (real) number nonpositive if it is absurd that it should be positive. Then Brouwer likewise claims that r is neither positive nor nonpositive[7], because if λ_f exists and is even, r is positive, if λ_f exists and is odd, r is negative, and if it does not exist, $r = 0$. Brouwer goes on to give further examples based on the same idea, some of them from analytic geometry. But I will note that the examples I have cited are all $\sum_1^0$ statements about a real number.

I do not want to dwell further on Brouwer, because intuitionism and its rejection of the law of excluded middle are well known, and going into Brouwer's underlying philosophy would take me too far afield. I will note that for many years I told philosophy of mathematics classes that Brouwer had

---

[4]In 1997 it was discovered by Yasumasa Kanada and Daisuke Takahasi that there *is* a sequence 0123456789 in the decimal expansion of π. See Jonathan M. Borwein, 'Brouwer-Heyting Sequences Converge', *Mathematical Intelligencer* 20, No. 1 (1998): 14–15. (Thanks to Mark van Atten for the information and references.) Brouwer's number k is between 17 and 18 billion. A less elementary example that Brouwer gives in the place cited above turns on whether there is an unpredictable number of such sequences in the expansion. That example is of course no longer $\sum_1^0$.

[5]'Mathematik, Wissenschaft, und Sprache', 161; trans, 51, slightly modified. The translation of the last sentence incorporates a handwritten correction by Brouwer; see Translator's Note b, 53, or note 11 to the text in *Collected Works*.

[6]Since being even and odd are decidable properties, I don't know why Brouwer formulates this in the negative way given here.

[7]The first of these claims does not depend on the parity-free character of *f*.

shown that there is a conflict between applying classical logic (in particular LEM) and a requirement that all proofs should be constructive. Although there can be differences of opinion about what counts as a constructive proof, this point is at least independent of Brouwer's idiosyncratic philosophy. And it is enough to yield the conclusion that the application of classical logic commits one to stepping outside a domain of mathematics that has its own attractiveness and has a certain amount of tradition behind it. In that sense, even simple consequences of the law of excluded middle are 'mathematically substantive'.

II. Weyl

In the 1920s, Weyl expressed a view of quantifiers in mathematics inspired by Brouwer's, and he gives vivid formulations that are echoed in remarks by Hilbert. The fullest presentation is in his striking essay of 1921, 'Über die neue Grundlagenkrise der Mathematik'.[8] Weyl begins this essay by sketching the critique of basic reasoning in analysis and construction of an 'arithmetic analysis' that he had presented in *Das Kontinuum* a few years earlier. There, and in the sketch, he accepted classical logic for statements involving quantification over natural numbers and statements involving quantification over a well-defined domain of 'properties' which can be defined by a limited list of construction principles. In the later terminology of Paul Bernays, the Weyl of *Das Kontinuum* is 'platonist' with respect to natural numbers and arithmetically definable properties of natural numbers.[9] But in the earlier work, he says that the use of existential propositions presupposes that the domain is 'a closed system of determinate objects existing in themselves'.[10] At that point, he did not have in view alternatives to classical logic. Shortly thereafter, he introduced the less-loaded term 'extensionally definite' (*umfangsdefinit*),[11] but the formulation of *Das Kontinuum* recurs in later writings, as we shall see. But after the encounter with Brouwer reflected in 'Grundlagenkrise', it is presented as a presupposition for the application of classical logic.

In the 1921 essay, Weyl notes that Brouwer did not agree that this presupposition is ever satisfied for an infinite domain and goes into the question whether his own earlier view or Brouwer's is correct. He considers the question whether there is a natural number with a certain decidable property P. He rejects as senseless the idea that such an assertion could be revealed as false through running through all natural numbers and not finding one possessing P.

---

[8]Quotations are in the translation in Mancosu, *From Brouwer to Hilbert*.
[9]Bernays, 'Sur le platonisme dans les mathématiques'. Bernays remarks (58) that Kronecker's famous aphorism that the natural numbers were created by God, while everything else in mathematics is the work of man, fits Weyl's view of 1918 better than Kronecker's own, which was closer to finitism.
[10]*Das Kontinuum*, 4, my translation.
[11]'Der *circulus vitiosus* in der heutigen Begründung der Analysis', 85.

Only the finding *that has actually occurred* of a determinate number with the property P can give a justification for the answer 'Yes', and — since I cannot run a test through all numbers — only the insight, that it lies in the essence of number to have the property $\neg P$, can give a justification for the answer 'No'; Even for God no other ground for decision is available. *But these two possibilities do not stand to one another as assertion and negation.*[12]

But Weyl finds himself thrown back to his earlier standpoint by the thought that if he runs through the numbers, either the process will break off with the discovery of a number possessing P or not:

... it is either so or not so, without change and wavering and without a third possibility. (ibid.)

However, Weyl is finally converted by a reflection on the nature of existential statements. An existential statement (*Satz*) is not a judgment in the proper sense that asserts a state of affairs (*Sachverhalt*). Existential states of affairs are 'an empty invention of logicians' (ibid.). The existential statement such as 'there is an even number' is a 'judgment abstract' obtained from a judgment about an instance such as '2 is an even number'. He then makes a striking analogy:

If knowledge is a precious treasure, then the judgment abstract is a piece of paper indicating the presence of a treasure, without revealing at which place. Its only value can be to drive me to look for the treasure. The piece of paper is worthless as long as it is not realized by an underlying actual judgment like '2 is an even number'.[13]

What then of general propositions? A general statement such as 'For every number m, $m + 1 = 1 + m$' is also not a real judgment but a 'general instruction for judgments' (*Anweisung auf Urteile*).[14] If one encounters an individual number, say 17, then following this 'instruction' one can arrive at genuine judgments such as $17 + 1 = 1 + 17$.

Weyl draws from this account of quantified statements the conclusion that it is

---

[12]'Über die neue Grundlagenkrise', 54, trans. 97, modified.

[13]Ibid., 54, trans. 97–98. Weyl uses the word 'judgment' (*Urteil*) with a meaning close to that of 'proposition' is contemporary usage. This was pointed out in Dirk van Dalen, 'Hermann Weyl's Intuitionistic Mathematics', 157 n. 11.

[14]Ibid., 55, trans. 98. The translation quotes the German phrase in parentheses but erroneously replaces *auf* by *für*. Føllesdal suggests that *für* would have been more appropriate. Curiously, in the original printing the word is misprinted as *quf*. Clearly *auf* was intended, and the text is corrected to *auf* in the reprints in *Selecta Hermann Weyl*, 225, and in *Gesammelte Abhandlungen*, II, 157.

... completely meaningless to negate this sort of statements. This means that it is quite impossible even to formulate an 'axiom of the excluded middle' for them.[15]

This view of quantification in mathematics is expressed with great confidence, and Weyl takes it to decide the issue between intuitionism on the model of Brouwer and the position of his own *Das Kontinuum*. The more standard approach to classical analysis is presumably already out of play because of the critique of impredicativity in the latter work.[16]

The view of quantifiers in mathematics just set forth is certainly more restrictive than Brouwer's. However, it is not much supported by argument. The account in 'Grundlagenkrise' is the fullest I have found in his writings. The distinction between an existential statement and an instance of it is partly put in metaphysical terms: For the true instance, there is a state of affairs, but there are no states of affairs corresponding to existential statements. More revealing is a remark in Weyl's 'treasure' metaphor: The 'only value' of the 'piece of paper' corresponding to the judgment abstract is that it drives him to seek an instance; the piece of paper is 'worthless' until such an instance is found. So, if we can attribute meaning to an existential statement, it must consist in the fact that it is verified when a true instance has been found. What is rejected (given the contrast with the earlier view expressed a couple of paragraphs back) is the idea that such an instance might exist, and thus the statement might be true, even if we do not or possibly even cannot find the instance. Weyl might well have had in the back of his mind the intention-fulfillment theory of meaning of Husserl's *Logical Investigations*. There seems also to be an implicit finitism in the reading of universal statements. A genuine judgment arises only when a case is presented.

How committed was Weyl to this understanding of quantifiers in mathematics? One might get a clue on this point from Weyl's later writings. The issue is discussed in almost the same words in a more purely philosophical and expository paper of 1925.[17] However, there is a subtle shift in that the point about quantifiers is put into a section discussing Brouwer's 'intuitive mathematics'. He continued to find Brouwer's theory of the continuum extremely attractive; he evidently thought it expressed better than any other the idea of the continuum as not reducible to a set of discrete elements but rather a 'continuous flux'. Weyl ends the section with the remark that 'with Brouwer, mathematics gains the highest intuitive clarity', but that 'full of pain, the mathematician sees the greatest part of his towering theories dissolve in fog'.[18]

---

[15]Ibid., 56, trans. 99.
[16]I have discussed this elsewhere; see 'Realism and the Debate on Impredicativity, 1917–44'.
[17]'Die heuitige Erkenntnislage in der Mathematik', 19, trans. 133.
[18]Ibid., 23–24, trans. 136.

A section covering the same ground occurs in the monograph *Philosophie der Mathematik und Naturwissenschaft*. There the view of quantifiers is stated much more briefly and, surprisingly, implicitly attributed to Brouwer. The concluding remark about Brouwer's intuitionist mathematics is expanded:

> Mathematics with Brouwer gains its highest intuitive clarity. He succeeds in developing the beginnings of analysis in a natural manner, all the time preserving the contact with intuition much more closely than had been done before. It cannot be denied, however, that in advancing to higher and more general theories the inapplicability of the simple laws of classical logic eventually results in an almost unbearable awkwardness. And the mathematician watches with pain the larger part of his towering edifice which he believed to have been built with concrete blocks dissolve into mist before his eyes.[19]

These observations would suggest that to the extent that he continued to hold the view of quantifiers expressed in 'Grundlagenkrise', he viewed it as part of the foundations of intuitionistic mathematics after Brouwer. In spite of the endorsement expressed in that paper, he never adopted intuitionism as a mathematical practice. Dirk van Dalen observes that he published only one piece of constructive mathematics, a proof of the fundamental theorem of algebra.[20] But Weyl was not a mathematician who expresses a philosophy when philosophy is being discussed and then unreflectively proceeds to practice mathematics in a way incompatible with that philosophy. Although Brouwer's theory of the continuum never lost a certain appeal for him, his philosophy did change in the 1920s.

One factor was undoubtedly the Hilbert program, which was presented in public only after the publication of 'Grundlagenkrise'. Expository papers from 1925 onwards contain a section on 'intuitive mathematics', basically intuitionism, and one on 'symbolic mathematics', at least originally focused on the Hilbert program. He read Hilbert's view in a formalist way. He says that Hilbert admits

> ... that the force of contentual thinking does not extend further than Brouwer claims, that it is not in a position to bear the 'transfinite' modes of inference in mathematics, that there is no justification for the transfinite

---

[19]Ibid., 44 [75], translation from the English version, *Philosophy of Mathematics and Natural Science*, 54. In the sections discussed here, the revisions for the English edition are local and do not affect the discussion or quotations here. For this reason, I have given in brackets page numbers from the 7th ed., which can be presumed to be the same as those in the first post-war German edition of 1966. In the translation *Urteil* is rendered as 'proposition', e.g. on p. 50, in accordance with van Dalen's point mentioned in note 15.

[20]'Hermann Weyl's Intuitionistic Mathematics', 163.

statements of mathematics as contentual, evident truths. What Hilbert wants to secure is not the *truth* but the *consistency* of the old analysis.[21]

Weyl regarded the Hilbert program as a significant development, a way of defending the classical mathematical practice in which he himself continued to engage. But he wrote that Hilbert 'succeeded in saving classical mathematics by a *radical reinterpretation of its meaning* without reducing its inventory'.[22]

Another factor was undoubtedly reflection on theoretical physics and the role of mathematics in it. Weyl himself had made significant contributions to mathematical physics during the time of the foundational writings we have been reviewing. Statements of theoretical physics 'certainly lack that feature which Brouwer demands of the propositions of mathematics, namely, that each should carry within itself its own intuitively comprehensible meaning'.[23] Weyl briefly asserts a holistic view associated with Duhem and later Quine: 'Rather, what is tested by confronting theoretical physics with experience is the system as a whole'.[24] Weyl speaks in places of 'symbolic construction of the world' and of mathematics as Hilbert understands it as 'bold theoretical construction.' The theme of symbolic construction recurs in later writings of Weyl.[25]

Nonetheless, it appears that Weyl never gave up a critical view of logic. This shows itself most clearly in an article of 1946.[26] In an exposition of first-order logic early in the paper, he says that the category (i.e. domain) over which the variables range is assumed to be 'a closed realm of things existing in themselves, or, as we shall briefly say, is *existential*'.[27] That is essentially the formulation of *Das Kontinuum*. At the beginning of a later section on Brouwer, Weyl writes:

> Brouwer made it clear, as I think beyond any doubt, that there is no evidence supporting the belief in the existential character of the totality of all natural numbers, and hence the principle of excluded middle in the form 'Either there is a number of the given property γ, or all numbers have the property $\neg\gamma$' is without foundation.[28]

---

[21]'Die heutige Erkenntnislage', 24, also *Philosophie der Mathematik und Naturwissenchaft*, 44 [76], my translation. *Philosophy of Mathematics and Natural Science* translates *inhaltlich* as 'intuitive' (55). The rendering 'contentual', following Jean van Heijenoort and Stefan Bauer-Mengelberg, is now well established. Although the remark translated here occurs in the same words in both places cited, the discussion of the Hilbert program in 'Die heutige Erkenntnislage' is less pointed and more inconclusive than in the slightly later publication.

[22]'Diskussionsbemerkungen zu dem zweiten Hilbertschen Vortrag', 87, trans. 483.

[23]*Philosophie der Mathematik und Naturwissenschaft*, 49 [83–84], trans. 61.

[24]Ibid. The same views are stated in 'Discussionsbemerkungen', 87–88, trans. 484.

[25]For example 'Wissenschaft als symbolische Konstruktion des Menschen'. On this theme see Bell and Korté, §4.5.8. This article is very informative about Weyl's work and thought in general.

[26]'Mathematics and Logic'.

[27]Ibid., 3.

[28]Ibid., 7.

Weyl then states very briefly the view of the quantifiers from 'Grundlagenkrise'. He goes on to say

> Brouwer opened our eyes and made us see how far classical mathematics ... goes beyond such statements as can claim real meaning and truth founded on evidence.[29]

In the 1920s, he had found a way to accept classical mathematics and logic while admitting the point that he here attributes to Brouwer. That appears still to have been his attitude in the 1940s.

III. Hilbert

To dissociate the criticism of logic from constructivism, I will turn to David Hilbert. Hilbert was always a robust defender of classical mathematics, including not only classical logic but set theory, without the sort of reservation frequently expressed by Weyl. It is hard to imagine that Hilbert, who famously wrote that 'no one shall be able to drive us from the paradise that Cantor created for us' would speak, as Weyl did, of the 'Fall and Original Sin of set theory'.[30] It is not clear to me how committed Hilbert was to a critical view of logic, but one can find expressions of it in his public lectures of the 1920s. Consider the following passage:

> But in mathematics these equivalences [of $\neg \forall x Ax$ and $\exists x \neg Ax$ and of $\neg \exists x Ax$ and $\forall x \neg Ax$] are customarily assumed, without further proof, to be valid for infinitely many individuals as well; and with this step we leave the domain of the finite and enter the domain of transfinite modes of inference. If we were consistently and blithely to apply to infinite totalities procedures that are admissible in the finite case, then we would open the floodgates of error. This is the same source of mistakes that we are familiar with from analysis. In analysis, we are allowed to extend theorems that are valid for finite sums and products to infinite sums and products only if a special investigation of convergence guarantees the inference; similarly, here we may not treat the infinite sums and products
>
> $A1 \wedge A2 \wedge A3 \ldots$
>
> $A1 \vee A2 \vee A3 \ldots$

---

[29]Ibid. Cf. also the remarks on Russell, Gödel, and Brouwer in *Philosophy of Mathematics and Natural Science*, 234.
[30]Hilbert, 'Über das Unendliche', 170, trans. 376; and Weyl, *Philosophy of Mathematics and Natural Science*, 234. Although Weyl used classical methods in mathematical work throughout his career, so far as I know he kept a distance from higher set theory.

As though they were finite, unless the proof theory we are about to discuss permits such a treatment.[31]

In the next paragraph, Hilbert goes on to express a somewhat skeptical view of the meaning of quantifiers in application to an infinite domain. The remarks may reflect the influence of Weyl or Brouwer; he begins by saying that the negation of a universal judgment 'has no precise content whatever'; it can obtain such a content if a counterexample is exhibited or if its assumption is shown to lead to a contradiction.

This essay was the first in which Hilbert presented a formalization of classical quantificational logic. As he did subsequently, he referred to the main axiom introduced for this purpose as the 'transfinite axiom'. He prefaces its introduction by remarking that for a rigorous grounding of mathematics 'we are not entitled to adopt as logically unproblematic the usual modes of inference that we find in analysis'.[32] It is clear that Hilbert means to include in this what we would call first-order logic.

Hilbert's most explicit later treatment of this issue, in the famous lecture on the infinite, is more directly tied to motivating or expressing the finitary standpoint of his proof theory. Hilbert discusses the Euclidean proof of the infinity of primes. It is quite in accord with the finitary standpoint to prove that for every prime p, there is a prime between $p + 1$ and $p! + 1$.

> ... and this leads us to formulate a proposition that expresses only a part of Euclid's proposition, namely: there exists a prime number that is $> p$. So far as content is concerned, this is a much weaker assertion, stating only a part of Euclid's proposition; nevertheless, no matter how harmless the transition appears to be, there is a leap into the transfinite when this partial proposition, taken out of the context above, is stated as an independent assertion.[33]

Hilbert then goes on to give explanations of general statements from a finitary point of view. These explanations echo rather directly the view of such statements presented by Weyl in 'Grundlagenkrise' and repeated in several later writings.

> In general, from the finitist point of view an existential proposition of the form 'There exists a number having this or that property' has meaning only as a partial proposition, that is, as part of a proposition that is more precisely determined but whose exact content is inessential for many applications. ... In like manner, we come upon a transfinite proposition

---

[31]'Die logischen Grundlagen der Mathematik', 155, trans. 1139–40.
[32]Ibid., 156, trans. 1140.
[33]'Über das Unendliche', 172, trans. 378.

when we negate a universal assertion, that is, one that extends to arbitrary numerals. So, for example, the proposition that, if a is a numeral, we must always have

$$a + 1 = 1 + a$$

is from the finitist point of view incapable of being negated.[34]

The universal proposition can from the finitist point of view 'only be interpreted ... as a hypothetical judgment that comes to assert something when a numeral is given' (ibid.). Thus, we cannot apply the disjunction according to which the equation just quoted is either satisfied for every numeral or is refuted by a counterexample. But there can be no question in mathematical practice of giving up either the laws of Aristotelian logic or negating arbitrary statements.[35]

Hilbert does not return to this issue in his public lectures. However, essentially the same view about universal and existential statements about numbers is expressed in more detail in the course of the explanation of the finitary standpoint in §1 of Hilbert and Bernays, *Grundlagen der Mathematik I*.[36]

IV. Is There a Better Argument for the Critical Review?

We have found in the literature we have discussed little argument for the view of logic that concerns me, particularly if it is to be a criticism of the application of classical logic in mathematics and not just a specification of the logic appropriate to a particular mathematical practice satisfying the exigencies of a foundational program like Hilbert's. The arguments either stated by our authors or implicit in our remarks assume either a verificationist view of mathematical truth, and an interpretation of the connectives in accord with that, or a requirement that the constituents of a proposition should be exhibited in intuition, whatever specifically is meant by intuition.

The constructivist critique of classical logic was an impressive achievement, because it was possible to build different versions of constructive mathematics on it. Brouwer and his constructivist successors were able to show that a coherent mathematical practice is possible that accommodates the criticism of classical logic that Brouwer pioneered. However, most mathematicians have not been willing to give up mathematics based on classical logic on the basis of either some first philosophy or a methodological requirement of constructivity. Constructive mathematics has its own interest and attraction, but it is a

---

[34] Ibid., 173, trans. 378.
[35] Ibid., 174.
[36] See 32–34, unchanged in 2d ed. (1968). Although this section was written by Bernays, it was almost certainly written at a time when Bernays was still in close contact with Hilbert.

minority pursuit. Moreover, the demand that the systems that formalize classical mathematics should be proved consistent by constructive methods has not been widely accepted. The impossibility of such a proof even for first-order arithmetic by Hilbert's finitary method is generally regarded as a consequence of Gödel's incompleteness theorem. Using constructive methods that go beyond even Brouwer's own version of intuitionism, consistency proofs and other proof-theoretic results have been obtained for subsystems of second-order arithmetic that are quite powerful, but it may be that any methods that we can now recognize to be constructive will be provably insufficient for going much further into second-order arithmetic than existing proof theory has gone.[37]

However, the examples of Hilbert and (with doubts and hesitations) Weyl show that a critical view of logic does not of itself imply the rejection of classical logic in mathematics. We can view the admission of quantifiers with infinite domains and the application of the law of excluded middle to statements involving them as substantive mathematical assumptions. Is there an argument for this view that does not depend on the sort of assumptions that underlie the philosophy of constructivism?

In this section, I will explore whether a critical view of logic gains support from what I call the entanglement of logic and mathematics. Elsewhere I discuss consequences of this phenomenon.[38] I may well have given the reader the impression that the thesis of entanglement can be neatly stated, in a couple of sentences. But what the reader of my paper will actually find is a series of illustrations, from first-order, second-order, and plural logic. The idea can be found in a classic paper of Paul Bernays,[39] but there too one does not find a succinct statement. One might think that at least first-order logic is quite independent of mathematics, because a valid or correctly derived formula has no ontological commitment except the trivial (and in principle avoidable) one to the nonemptiness of the domain. But even very simple reflection on logic leads to mathematical commitments, already if one goes beyond deriving particular formulae or inferences to claims about what is derivable. A statement of derivability is a statement that *there is* a derivation. Conversely, a claim of nonderivability of a formula A is a universal statement, to the effect that no derivation is a derivation of A. Even in first-order logic, claims that particular formulae are not valid, or are satisfiable, can be more complex. As we know, in some cases, nonvalidity or satisfiability can only be witnessed by structures with a countably infinite domain, with relations that are not decidable.[40] Even in monadic logic, although any nonvalid formula has a finite countermodel, for formulae, generally the minimal size of such a model increases with the

---

[37]Per Martin-Löf argues to this effect in 'The Hilbert-Brouwer Controversy Resolved?'. Michael Rathjen suggests a bound higher than Martin-Löf's but still within second-order arithmetic in 'The Constructive Hilbert Program', §6.
[38]'Some Consequences of the Entanglement of Logic and Mathematics'.
[39]'Die Philosophie der Mathematik und die Hilbertsche Beweistheorie'. See especially part I §2.
[40]For further discussion see 'Some Consequences', section I.

complexity of the formula. One might ask why traditional logic was able for so many centuries to stay aloof from mathematics. One reason is surely that if an inference of the form of a syllogism is not valid, a counterexample can be found with a domain of one or two elements. Furthermore, the total number of valid syllogisms is quite manageable finite number.

I will begin with a relatively easy case, second-order logic. Second-order logic should be considered from two points of view, model theoretically and as a deductive instrument. If we leave aside formulations restricted to insure predicativity, the most natural model theory for second-order logic is what is called the standard one, where, if the individuals range over a certain domain D of objects, the second-order variables range over all subsets of D (for the monadic case) and all n-ary relations of objects in D (for the polyadic case).[41] Neither the existing deductive systems for second-order logic nor recursively axiomatizable extensions of them are complete for this model theory.

But the most problematic aspect of second-order logic from this point of view is that questions of validity can be equivalent to questions of truth in set theory, where the set-theoretic propositions themselves can be independent of widely accepted axioms. As has been noted by many writers, the truth of the continuum hypothesis is equivalent to the validity in standard models of a second-order formula. The real numbers can be characterized up to isomorphism by a single second-order formula; call it R. CH, with reference to that structure, would say that there is no subset neither equinumerous to a subset of the natural numbers nor equinumerous to the whole domain. Assuming R satisfiable (which is provable in Zermelo set theory), R $\rightarrow$ CH is valid if and only if the continuum hypothesis is true as a set-theoretic proposition, and R $\rightarrow$ ¬CH is valid if and only if it is false. It follows that standard second-order validity can be upset by forcing.

Axiomatic second-order logic as formulated in standard works does not differ in its deductive power from a first-order theory of a domain of individuals and its subsets and relations. Although it is complete with respect to Henkin models, there are ways of extending it, for example, by going to third-order logic, but since second-order arithmetic can be axiomatized in this language, Gödel's incompleteness theorem applies even if we ascend to third-order and higher order logic. Eventually, we are led into set theory.[42]

It is a staple of older debates on foundations that full second-order logic applied in arithmetic introduces the issue of impredicativity. We have this even if the second-order logic is intuitionistic. A common view of the classical case is that it imposes a realistic understanding of the entities (which we can assume to be sets) that the second-order variables range over. At any rate, we cannot

---

[41] Intensional versions can be proposed where the second-order variables range over arbitrary attributes and relations in intension of the objects in D. The clarity of these conceptions can be questioned, but they can be modeled mathematically by possible-worlds semantics. That understanding is parasitic on the set-theoretic interpretation. I leave it out of account here.

[42] This point is argued in more detail in 'Some Consequence', Section III.

require of a set that it be definable in a way that only involves reference to or quantification over sets 'constructed earlier'. Whatever our conception of sets is that licenses impredicativity, admitting them is a substantive mathematical assumption. This is an assumption that most mathematicians will accept, although the work in reverse mathematics shows that it is needed for less of mathematical practice than one might think. We could express the assumption in terms derived from Weyl, that for a given domain of individuals, which we will suppose extensionally definite in his sense, the subsets of that domain also comprise an extensionally definite domain. If the domain of individuals is the natural numbers, the resulting theory, second-order arithmetic, is already a very powerful theory. We have seen that if the domain of individuals is the real numbers, we can formulate the continuum hypothesis, a proposition where it is at present not agreed that it has a determinate truth value.

I would infer from this that second-order logic is not only entangled with set theory, tending to justify W V Quine's memorable characterization of it as 'set theory in sheep's clothing'.[43] It is entangled with a rather problematic aspect of set theory, its posing of problems that there are principled obstacles to answering.[44] Should we then accept first-order logic as the basic logic for reasoning about objects generally, maybe even as self-evident?[45] The fact that Brouwer showed that applying classical first-order logic leads to the acceptance of non-constructive proofs might give us pause.

Brouwer and other early critics stressed the application of logic in reasoning about the infinite. In analysis, however, logic is applied to reasoning about the uncountable infinite, and in higher set theory, about ever larger and higher infinite totalities. To be sure, if something goes wrong, we can always 'blame' it on mathematical assumptions that do not belong to the logic. If a large cardinal axiom proves to be inconsistent, set theorists scrap that axiom and do not change the logic. That is what happened some years ago in the case of a 'Reinhardt cardinal', that is, a cardinal κ such that it is the critical point of an elementary embedding $j: V \to V$ that is not the identity.[46] The existence of such a cardinal was proved by Kenneth Kunen to be inconsistent with ZFC.[47]

---

[43] *Philosophy of Logic*, 66 (either ed.). Quine's arguments for this view are interesting but not the most powerful ones, which rely on the serious entanglement of second-order logic and set theory sketched above. For deeper remarks on this theme, see Väänänen, 'Second-Order Logic and Foundations of Mathematics'; and Koellner, 'Strong Logics of First and Second Order'.
[44] I don't mean to argue that second-order logic is not useful or important; after all who would deny that set theory has these properties? I don't even mean to claim that there is no reasonable sense in which it can be called logic. However, I have argued in other writings that it has been oversold as an instrument for recent programs in the philosophy of mathematics.
[45] Cf. Føllesdal, 'Comment on Parsons'.
[46] The critical point would be the first ordinal that is moved by the embedding. See Akihiro Kanamori, *The Higher Infinite*, §5.
[47] See Kanamori, op. cit., 318–19. There is a complication because Kunen's proof relied on the axiom of choice. It is still not known whether the existence of a Reinhardt cardinal is inconsistent with ZF without Choice.

Furthermore, if one does not trust first-order logic in reasoning from assumptions whose consistency one might doubt, how can we trust the logic used in deriving an inconsistency such as the one that resulted from the idea of a Reinhardt cardinal? Of course, the inconsistency result is a syntactic result, namely that a certain deduction from axioms results in a contradiction. This does not depend on the soundness of the logic. However, the natural response, so far as I know shared by all set theorists, is that the fault lies with the assumption of a Reinhardt cardinal and not with the logic.

It still remains that our logical intuitions (even for second-order logic, which dates from the 1870s) were formed from applications involving infinities not higher than those of analysis, and yet they are applied in set theory to the highest infinities we are able to describe precisely so far, and also to reasoning involving statements that those axioms that are widely accepted are known not to settle.

The basic logical intuitions underlying first-order logic could be divided into those about inference and those about truth. Attempts to vindicate first-order logic on the basis of purely inferential considerations have generally aimed at intuitionistic logic, and it has seemed to me that extension to classical logic has required going beyond the strictly inferential standpoint. On the other hand, intuitions about truth lead rather immediately to the law of excluded middle.

However, intuitionistic logic has not been significant in practice once one ventures into set-theoretic mathematics. In logically weaker contexts, such as first-order arithmetic, it has been known for some time that there is a translation of classical theories into intuitionist and that a classical theory is a conservative extension of the corresponding intuitionistic one for a significant class of formulae. A similar translation exists for second-order arithmetic, but second-order arithmetic with intuitionistic logic has not generally been accepted as constructive, so that little is gained by it. The set theories that are significant for current constructive mathematics, such as Peter Aczel's CZF, are in terms of interpretability much weaker than classical theories such as ZF, even though ZF results from CZF when the law of excluded middle is added.[48]

Any argument for a critical view of logic based on such considerations will be in the same circle we encountered earlier, of depending on considerations of constructivity. I want instead to follow out a little further the suggestion made above that our logical intuitions were formed in connection with mathematics involving quite low levels of infinity. How evident is it that even first-order logic is correct in application to the higher infinite?

Let us first look at the soundness proofs for first-order logic. The soundness of standard proof systems can be proved in very weak theories. There is a limit to what such proofs can do for us, for the obvious reason that they proceed by reasoning of the same type as that which the logic whose soundness is at issue

---

[48] See Aczel and Rathjen, *Notes on Constructive Set Theory*, proposition 7.5.

formalizes. But there is another limitation of such proofs, if we consider their application to a piece of reasoning in set theory. Let's suppose that a certain large cardinal axiom A has been used, in addition to ZFC, to prove a certain theorem T. Formalizing the proof will yield the result that the formula

$$(1)\ \text{ZFC}_m \wedge A \rightarrow T,$$

is derivable, where ZFC_m is a finite conjunction of axioms of ZFC, those needed for the formal proof. The soundness of the proof system for first-order logic implies that (1) is valid. But the relatively weak systems in which that can be proved do not rule out the possibility that this validity is vacuous, in other words that it obtains because the antecedent is not satisfiable, so that (1) is made true by falsity of antecedent. There is nothing surprising in this: any proof that insured that nonvacuous validity of (1) would imply the consistency of $\text{ZFC}_m + A$, and, since A will typically be strongly inaccessible, probably the consistency of $\text{ZFC} + A$.

The fact that our logical intuitions were formed before we had any significant experience in reasoning with the higher infinite gets a little reinforcement from the observation that arguments for the soundness of first-order logic do not engage the higher infinite and, in the absence of very strong premises, are compatible with the falsity of strong set theories. On the other hand, they are quite general, as is illustrated by the fact that theories in which soundness is proved can speak of quite arbitrary sets. But what we know about the scope of such generalization is dependent on the very development I have been reflecting on, the development of theories of the higher infinite relying on the same classical logic. We are left with a question: Is the evidence of logic applied to the higher infinite really independent of the evidence of the mathematical coherence of the higher infinite itself? The issue might be better viewed as one of understanding: How well do we understand logical validity in the context of reasoning about the higher infinite? Do we really understand it better than we understand the higher infinite itself?

Acknowledgements

I am grateful to Dagfinn Føllesdal for his extensive comments on an earlier version. A later version was presented to the conference on History and Philosophy of Infinity at Corpus Christi College, Cambridge, September 20, 2013. Thanks to Haim Gaifman for comments at the conference and to Mark van Atten for written comments on that version.

References

Aczel, Peter, Michael Rathjen. *Notes on Constructive Set Theory*. Report No. 40. Institut Mittag-Leffler, 2000–2001.

Bell, John L., Herbert Korté. 'Hermann Weyl'. *Stanford Encyclopedia of Philosophy.* 2009, revised 2011. http://plato.stanford.edu/entries/weyl/

Bernays, Paul. 'Die Philosophie der Mathematik und die Hilbertsche Beweistheorie'. *Blätter für deutsche Philosophie* 4 (1930): 326–67. (Reprinted with postscript in *Abhandlungen zur Philosophie der Mathematik.* Darmstadt: Wissenschaftliche Buchgesellschaft, 1976.)

Bernays, Paul. 'Sur le platonisme dans les mathématiques'. *L'enseignement mathématique* 52 (1935): 52–69.

Borwein, Jonathan M. 'Brouwer-Heyting Sequences Converge'. *The Mathematical Intelligencer* 20, no. 1 (1998): 14–5.

Brouwer, L.E.J. *Collected Works*, vol. 1: *Philosophy and Intuitionistic Mathematics*, ed. A. Heyting. Amsterdam: North-Holland, 1975.

Brouwer, L.E.J. 'Mathematik, Wissenschaft, und Sprache'. *Monatshefte für Mathematik und Physik* 36 (1929): 153–64. (Translation in Mancosu, *From Brouwer to Hilbert.*)

Brouwer, L.E.J. 'Über die Bedeutung des Satzes vom ausgeschlossenen Dritten in der Mathematik, insbesondere in der Funktionentheorie'. *Journal für die reine und angewandte Mathematik* 154 (1924): 1–7. (Translation in Heijenoort, *From Frege to Gödel.*)

Burge, Tyler. 'Frege on Knowing the Foundation'. *Mind* 107 (1998): 305–47. (Reprinted in *Truth, Thought, Reason: Essays on Frege.* Oxford: Clarendon Press, 2005.)

Dalen, Dirk van. 'Hermann Weyl's Intuitionistic Mathematics'. *The Bulletin of Symbolic Logic* 1 (1995): 145–69.

Føllesdal, Dagfinn. 'Comment on Parsons'. In *Reference, Rationality, and Phenomenology*, ed. M. Frauchiger, 303–06. Frankfurt: Ontos Verlag, 2013.

Frauchiger, Michael, ed. *Reference, Rationality, and Phenomenology: Themes from Føllesdal.* Frankfurt: Ontos Verlag, 2013.

Heijenoort, Jean van, ed. *From Frege to Gödel: A Source Book in Mathematical Logic, 1879–1931.* Cambridge, MA: Harvard University Press, 1967.

Hilbert, David. 'Die logischen Grundlagen der Mathematik'. *Mathematische Annalen* 88 (1923): 151–65. (Translation in Ewald, W., ed. *From Kant to Hilbert.* Oxford: Clarendon Press, 1996, volume II.)

Hilbert, David. 'Über das Unendliche'. *Mathematische Annalen* 95 (1926): 161–90. (Translation in van Heijenoort, *From Frege to Gödel.*)

Hilbert, David, Paul Bernays. *Grundlagen der Mathematik I.* Berlin: Springer, 1934. 2d ed., 1968.

Kanamori, Akihiro. *The Higher Infinite.* Berlin: Springer-Verlag, 1994. 2d ed., 2003.

Koellner, Peter. 'Strong Logics of First and Second Order'. *The Bulletin of Symbolic Logic* 16 (2010): 1–36.

Mancosu, Paolo, ed. *From Brouwer to Hilbert: The Debate on the Foundations of Mathematics in the 1920s.* New York and Oxford: Oxford University Press, 1998.

Martin-Löf, Per. 'The Hilbert-Brouwer Controversy Resolved?'. In *One Hundred Years of Intuitionism (1907–2007)*, ed. Mark van Atten, Pascal Boldini, Michel Bourdeau, and Gerhard Heinzmann, 243–56. Basel: Birkhäuser, 2008.

Parsons, Charles. 'Realism and the Debate on Impredicativity, 1917–1944'. In *Reflections on the Foundations of Mathematics: Essays in Honor of Solomon Feferman*, ed. Wilfried Sieg, Richard Sommer, and Carolyn Talcott. Lecture Notes in Logic 15, 372–89. Natick, MA: Association for Symbolic Logic and A. K. Peters, 2002.

Parsons, Charles. 'Some Consequences of the Entanglement of Logic and Mathematics'. In *Reference, Rationality, and Phenomenology*, ed. M. Frauchiger, 153–78. Frankfurt: Ontos Verlag, 2013.

Quine, W.V. *Philosophy of Logic.* Englewood Cliffs, NJ: Prentice-Hall, 1970. 2d ed. Cambridge, MA: Harvard University Press, 1986

Rathjen, Michael. 'The Constructive Hilbert Program and the Limits of Martin-Löf Type Theory'. *Synthese* 147 (2005): 81–120.
Väänänen, Jouko. 'Second-Order Logic and Foundations of Mathematics'. *The Bulletin of Symbolic Logic* 7 (2001): 504–20.
Weyl, Hermann. 'Der *circulus vitiosus* in der heutigen Begründung der Analysis'. *Jahresbericht der Deutschen Mathematiker-Vereinigung* 28 (1919): 85–92.
Weyl, Hermann. 'Disknssionsbemerkungen zu dem zweiten Hilbertschen Vortrag über die Grundlagen der Mathematik'. *Abhandlungen* aus dem mathematischen *Seminar der Universität Hamburg* 6 (1928): 86–8. (Translation in Heijenoort, *From Frege to Gödel*.)
Weyl, Hermann. *Gesammelte Abhandlungen*, 4 vols. ed. K. Chandrasekaran. Berlin: Springer-Verlag, 1968.
Weyl, Hermann. 'Die heuitige Erkenntnislage in Der Mathematik'. *Symposion* 1 (1925): 1–32. Translation in Mancosu, *From Brouwer to Hilbert*.
Weyl, Hermann. *Das Kontinuum. Kritische Untersuchungen über die Grundlagen der Analysis*. Leipzig: Veit, 1918.
Weyl, Hermann. 'Mathematics and Logic'. *The American Mathematical Monthly* 53 (1946): 2–13.
Weyl, Hermann. *Philosophie der Mathematik und Naturwissenschaft*. Munich: Oldenbourg, 1927. 7th ed. 2000. (The first edition was part of volume 2 of a *Handbuch der Philosophie*, but the individual articles were paginated individually and were apparently also sold separately.)
Weyl, Hermann. *Philosophy of Mathematics and Natural Science*. Princeton, NJ: Princeton University Press, 1949. Translation, with revisions and additions, of *Philosophie der Mathematik und Naturwissenschaft*.
Weyl, Hermann. *Selecta Hermann Weyl*. Basel: Birkhäuser, 1956.
Weyl, Hermann. 'Über die neue Grundlagenkrise der Mathematik'. *Mathematische Zeitschrift* 10 (1921): 39–79. (Translation in Mancosu, *From Brouwer to Hilbert*.)
Weyl, Hermann. 'Wissenschaft als symbolische Konstruktion des Menschen'. *Eranos-Jahrbuch* (1948): 375–31. Reprinted in *Gesammelte Abhandlungen*, IV, 289–345.

Dummett's objection to the ontological route to intuitionistic logic: a rejoinder

Mark van Atten

ABSTRACT
In 'The philosophical basis of intuitionistic logic', Michael Dummett discusses two routes towards accepting intuitionistic rather than classical logic in number theory, one meaning-theoretical (his own) and the other ontological (Brouwer and Heyting's). He concludes that the former route is open, but the latter is closed. I reconstruct Dummett's argument against the ontological route and argue that it fails. Call a procedure 'investigative' if that in virtue of which a true proposition stating its outcome is true exists prior to the execution of that procedure; and 'generative' if the existence of that in virtue of which a true proposition stating its outcome is true is brought about by the execution of that procedure. The problem with Dummett's argument then is that a particular step in it, while correct for investigative procedures, is not correct for generative ones. But it is the latter that the ontological route is concerned with.

1. Introduction

Michael Dummett opens his paper 'The philosophical basis of intuitionistic logic'[1] by asking

> What plausible rationale can there be for repudiating, within mathematical reasoning, the canons of classical logic in favour of those of intuitionistic logic? (Dummett 1978a, 215)

and devotes the remainder to a discussion of two different rationales or routes that would lead to such a repudiation for the case of number theory, a meaning-theoretical one and an ontological one. The meaning-theoretical route starts from the general slogan that meaning

---
[1]That paper has its origin in an invited lecture, titled 'Philosophical foundations of intuitionistic logic', at the Logic Colloquium held in Bristol in 1973 (Rose and Sheperdson 1975, preface); the reworked text appeared in its proceedings (Dummett 1975, 5–40). I will refer to the reprint in Dummett (1978b, 215–247).

is use. Logic is part of our language, a language that each of us has to be able to learn from our fellow human beings' use of it, and which we can only be said to use correctly if we can make our understanding manifest. Dummett argues that classical logic does not meet these two conditions, but intuitionistic logic does. The ontological route, on the other hand, starts from the thesis that mathematical objects are mental constructions, and as such come into being at some point, and proceeds to the conclusion that the appropriate logic for statements about such objects is not classical but intuitionistic. Of these two routes to intuitionistic logic, Dummett argues that the meaning-theoretical one may turn out to be correct if pursued, but the ontological route must be rejected. The positive proposal about the meaning-theoretical route, along which Dummett himself was already travelling[2] and would continue to for the rest of his career, has generated much discussion, no doubt because the underlying idea that meaning is use is not limited to mathematics or any other specific domain. On the other hand, the negative proposal about the ontological route, on which Dummett never came to have second thoughts,[3] has received much less comment.[4]

---

[2]Dummett (2007a, 17–18) traces his meaning-theoretical turn to the final part of his essay 'Truth' (Dummett 1959).

[3]In both the 1977 and 2000 edition of *Elements of Intuitionism* the argument is repeated, but then 'it is left to the reader to think through' the question 'to what extent such a radical anti-realism with respect to the objects of mathematics is defensible' (Dummett 1977, 389; 2000, 269). However, Dummett's own mind was made up. In the preface to *Truth and Other Enigmas* of 1978, he singles out 'The philosophical basis of intuitionistic logic' as one of the two essays with which 'I should, for the time being, feel completely satisfied' (Dummett 1978b, x–xi). In his book on Frege's philosophy of mathematics of 1991 he writes:

> I have argued in various places that the only route to a vindication or refutation of realism must go through a meaning-theoretic enquiry into the right form of semantic theory. (Dummett 1991a, 198n.13)

which implies a continued rejection of the ontological route to intuitionistic logic, and Dummett uses the argument to the same effect in 1994 in his 'Reply to Prawitz' (Dummett 1994, 296–297). It appears again, with fewer comments, in 2007 in his 'Reply to Künne' (Dummett 2007b, 349). One would have expected Dummett to bring up the matter, and elaborately at that, in his review (Dummett 1980) of volume 1 of Brouwer's *Collected Works* (Brouwer 1975). That was Dummett's first and only direct engagement with Brouwer's foundational writings in print, even though, as he told Joachim Schulte in 1987, he 'began to read all those works of Brouwer that are such heavy going' in 1962 (Dummett 1993a, 175).

[4]I mention here passages in Wright and Placek. Wright attempts to show that the form of intuitionism in arithmetic that Dummett finds acceptable – the form in which Argument A of Section 2 below is accepted – collapses into 'arithmetical Platonism' (Wright 1982, 245–246), which of course is just as inconsistent with the ontological intuitionism Dummett rejects; but ontological intuitionism itself is not discussed by Wright. Likewise, in his book he takes it as given that mathematical procedures are investigative, not generative; see footnote 20 below. Placek (1999) discusses the analogy Dummett draws, in *Elements of Intuitionism* (Dummett 1977, 384–386; 2000, 265–267) between statements about the natural numbers and statements about fictional characters in literature when both are read as statements about mental creations. However, he stops short of considering Argument A, presumably because he considers the idea that 'bivalence must already fail for the arithmetical statements

Dummett's objection to the ontological route is important for two reasons:

(1) He took this objection to show that a meaning-theoretical route is the only viable approach to the adoption of intuitionistic logic in mathematics.
(2) He had already objected to the approach of Brouwer and Heyting on account of what he took to be its psychologism (Dummett 1963), which is one version of the ontological route; but the present objection is much more general, having as scope any defence of intuitionistic logic from an ontological premiss, thus including non-psychologistic but still ontological approaches.[5]

Dummett presents his objection to the ontological route as part of an analysis of the dispute between the realist and the intuitionist, its exact content, and the conditions under which it can be resolved. I will say something on that below, but in passing, as the objection can be made independently, and threatens the position also of the intuitionist who is not overly concerned with the debate with realists.

2. The ontological route and Dummett's objection to it

The ontological intuitionist starts from this thesis:

> T1 Natural numbers are the creations of human thought. (Dummett 1978a, 228–229)

and relates it to truth by

> T2 A true mathematical statement is true in virtue of a proof having actually been effected which justifies that statement (Dummett 1978a, 247).[6]

T2 is an instantiation of Dummett's Principle C:

---

of which it has not actually been decided whether or not they hold, although the relevant decision procedures are known' to be 'a position completely alien to the intuitionist' (Placek 1999, 123–125).

[5] For example, transcendental-idealist readings in the sense of Kant or Husserl; see, e.g. Posy (1991) and chapter 6 of van Atten (2004). Dummett has written that 'a non-psychologistic form of idealism is possible, and indeed the only version worth considering' (Dummett 1981, 68).

[6] Other forms of constructivism hold that a true mathematical statement is true in virtue of the existence of a constructive proof that need not even be known; this involves a notion of existence that Brouwer and Heyting reject. A philosophical development of such a view is found in the work of Dag Prawitz for example (Prawitz 1998b); I choose this particular reference because Dummett commented on it at length and was influenced by it (Dummett 1998a, 1998b; and Pataut 2002, 244–245). See also page 9 below.

If a statement is true, there must be something in virtue of which it is true. (Dummett 1959, 14; 1976, 52; 1973, 464)[7]

Künne has introduced the distinction between propositional readings of Principle C, according to which this something is the truth of some proposition, and ontic readings, according to which it is the existence of something else (Künne 2007, 321). Dummett, in his reply to Künne, prefers propositional readings (Dummett 2007b, 346). So when Heyting writes that

> A mathematical assertion affirms the fact that a certain mathematical construction has been effected. (Heyting 1956, 3)

I take it that Dummett would read this propositionally by bringing to bear on this his Fregean view that facts are true propositions and not objects of a different ontic category (Dummett 2007b, 346). But Heyting surely meant this ontically, as would Brouwer; for them mathematics consists essentially in languageless constructions based on the intuition of time. Their notion of proof, in its most fundamental sense, is identified with that of such constructions:

> If mathematics consists of mental constructions, then every mathematical theorem is the expression of a result of a successful construction. The proof of the theorem consists in this construction itself, and the steps of the proof are the same as the steps of the mathematical construction. (Heyting 1958, 107)

Dummett's objection to ontological intuitionism, however, will not turn on the difference between these two readings of Principle C, but on the fact that the ontological intuitionist instantiates Principle C in such a way that that in virtue of which a true mathematical statement is true is something that comes into being at some point.

T2 immediately leads to intuitionistic logic because of its verificationist nature. It might seem that T1 suffices for this because it would seem to entail T2. However, perhaps it could be held that, before the creation of numbers, the propositions about what will be the case when they have been created already have their truth values independently of our knowledge, which would motivate accepting T1 together with classical logic. Dummett names Dedekind as someone who may have held that position

---

[7] It lies to hand to use the term 'truth-maker' here (e.g. Sundholm 1994, 123–124n.20); Dummett, on the other hand, has written that 'I am allergic to the notion of "truthmakers", and do not much like to talk about what makes a statement true, rather than that in virtue of which it is true' (Dummett 2007b, 346). For my present purpose, nothing hinges on this.

(Dummett 1978a, 246). In his discussion of the ontological route, he wishes to leave this possibility open.

In his argument against ontological intuitionism, Dummett's target is therefore not the ontological thesis by itself. It is, as we will see, rather this consequence of T2:

> T2-d A *decidable* true mathematical statement is true in virtue of our actually having effected a proof which justifies the statement.

In the context of the debate between ontological intuitionists and Platonists, the importance of T2-d is that it concerns what Dummett considers to be the common ground between their positions, without which there could properly speaking be no dispute: both sides agree that understanding a decidable statement consists in mastering a decision procedure for it (Dummett 1978a, 238). Dummett holds that in that case there is, in spite of the different accounts that the Platonist and the intuitionist give of how a decision procedure that we master relates to the truth value of the statement it decides, no serious disagreement over the meaning of these statements. This leaves open the possibility that an argument can be found that allows extending the intuitionistic view on that relation to a view on the meaning of undecidable statements as well. If it can, that would be a way for the intuitionist to make his position understandable to the Platonist, which would, in turn, be a condition for the possibility of eventually convincing the latter.

Dummett's objection to the ontological route will not be that there is a logical or conceptual problem with T2-d itself, but that accepting it obliges one also to accept a form of scepticism that he considers too high a price. I will present his argument (Dummett 1978a, 244–247) as consisting of two sub-arguments that I call A and B:

> Argument A seeks to establish that decidable statements have their truth values already before a corresponding decision procedure is carried out.
>
> Argument B leads from an observation about the structure of Argument A to the rejection of ontological intuitionism.

Dummett's reasoning that I reconstruct as Argument A runs as follows. Let n be a natural number, P a decidable predicate defined on the natural numbers, and D a decision procedure for propositions of the form $P(n)$.

A1 If, for a given n, we were to carry out the decision procedure D, then we should find that $P(n)$ holds, or that $\neg P(n)$ holds. (Premiss)

A2 The decision procedure D needs no further information, and does not depend on chance. (Premiss)

A3 If, for a given n, we were to carry out the decision procedure D, then we should find that $P(n)$ holds, or, if we were to carry out the decision procedure D, then we should find that $\neg P(n)$ holds. (From A1 and A2)

A4 For a given n, the statement $P(n)$ already has a determinate truth-value prior to the execution of the decision procedure D. (From A3)[8]

Premiss A1, which Dummett takes as given, states that for decidable predicates the principle of the excluded middle holds; the conclusion A4 that for these predicates also the principle of bivalence holds. Dummett states that

> It is very difficult for us to resist the temptation to suppose that there is already, unknown to us, a determinate answer to the question which of the two disjuncts we should obtain a proof of, were we to apply the decision procedure. (Dummett 1978a, 244)

But, he observes, it would not be possible to infer directly from A1 to A3 as an application of the schema

$$\frac{A \to (B \vee C)}{(A \to B) \vee (A \to C)}$$

giving '$\to$' the meaning of the subjunctive conditional in natural language,[9] and instantiating A with 'We carry out decision procedure D for $P(n)$', B with 'We find that $P(n)$', and C with 'We find that $\neg P(n)$'. He identifies 'two obvious kinds of counter-example' (Dummett 1978a, 244–245) to the schema. In the first kind, whether it is B or rather C that would be true is determined not by A alone but by A and an as yet unspecified further circumstance Q, such that $(A \wedge Q) \to B$ and $(A \wedge \neg Q) \to C$. To take Dummett's example of 1978, if Fidel Castro were to meet President Carter, he would either insult him or be polite, but which of the two might depend on whether they met in Cuba or elsewhere.[10] In the second kind, whether it is B or rather C that would be true if A were depends solely on chance. For instance, A may refer to some quantum-mechanical event whose possible outcomes lead to B and C, respectively. Dummett suspects

---

[8] A direct consequence of A4 is that the logic of decidable statements is classical; this means also that, if Dummett is right, to accept that logic for such statements the Platonist does not need his ontological thesis.

[9] Dummett writes '$\to$' for both intuitionistic and subjunctive conditionals, specifying its interpretation as required.

[10] In the 1975 publication based on the 1973 lecture, the example of course involved Nixon (Dummett 1975, 37).

both kinds of counterexample are subtypes of a single type (Dummett 1978a, 244); this seems right, in that both turn on the fact that the information that would be required to accept the consequent cannot be recovered from that contained in the antecedent.

Once premiss A2 is added, which is likewise taken as given, obviously no counter-examples of these two types can be generated; and as Dummett holds that it is hard to see on what other grounds the inference from A1 alone to A3 might fail, he concludes that the inference from A1 and A2 to A3 is acceptable. That inference is not deductive; but because it is an inference about procedures that are mathematical, Dummett is in this case prepared squarely to locate 'the difficulty of resisting the conclusion' (Dummett 1978a, 245) in the absence of any other 'obvious kind of counter-example' (Dummett 1978a, 244):

> No further circumstance could be relevant to the result of the procedure – this is part of what is meant by calling it a computation; and, since at each step the outcome of the procedure is determined, *how can we deny* that the overall outcome is determinate also? (Dummett 1978a, 245, emphasis mine)

Dummett's move from A3 to A4 is immediate. After the above passage, he continues:

> If we yield to this line of thought, then we must hold that every statement formed by applying a decidable predicate to a specific natural number already has a definite truth-value, true or false, although we may not know it. (Dummett 1978a, 245)

Note also how the choice of the verb 'yield' here again signals the non-deductive character of the inference. Likewise, in the later presentation in his 'Reply to Prawitz' he writes:

> The intuitionist sanctions the assertion, for any natural number, however large, that it is either prime or composite, since we have a method that will, at least in principle, decide the question. But suppose that we do not, and in practice cannot, apply that method: is there nevertheless a fact of the matter concerning whether the number is prime or not? *There is a strong impulse to say that there must be:* for surely there must be a definite answer to the question what we should get, were we to apply our decision method. (Dummett 1994, 296–297, emphasis mine)

And in 2007 he writes of this inference from the holding of the principle of the excluded middle to that of bivalence that he is '*disposed to think* that, in an instance of this kind, it does follow in the mathematical case, but not in the empirical case' (Dummett 2007b, 349, emphasis mine).

While the conjunction of A1 and A2 does not imply A3, A3 does imply the conjunction of A1 and A2. In that implication, which is not an instance of Ex Falso Sequitur Quodlibet, A3 can be said to *explain* why A1 and A2 are true: it would be *because* these procedures have the property ascribed to them in A3 that they have the properties ascribed to them in A1 and A2. That conception of the role of A3 corresponds to this remark in *Elements of Intuitionism*:

> The cases in which we *are* prepared to allow the inference from $P \to Q \vee R$ to $(P \to Q) \vee (P \to R)$ are precisely those in which we do believe in the existence of an objective reality which determines one or other conditional as true. (Dummett 2000, 268, original emphasis)

This applies to the case at hand, because we take A2 further to specify our understanding of A1 in such a way as to allow the inference to A3.

Dummett does not attempt to describe the range of alternative explanations. The reason for this seems to be that, to be essentially different, they should involve a refusal to accept the inference from A1 and A2 to A3; but Dummett considers that inference to be 'a line of thought that is *overpoweringly natural* for us' (Dummett 1994, 297, emphasis mine), and says that denying its acceptability amounts to a 'resolute' and 'hardheaded' scepticism concerning subjunctive conditionals (Dummett 1978a, 247).[11]

We have seen, then, that

(1) the inference from A1 and A2 to A3 is not deductive, and neither is it taken to be;
(2) A3 would explain the joint truth of A1 and A2;
(3) and, in Dummett's view, no alternative would explain this as well as A3 would, as any alternative requires us to refuse accepting a subjunctive conditional that we find highly natural to accept.

---

[11]Dummett acknowledged that he could take philosophical stances that he considered hard-headed himself. Of his own acceptance of the claim, made by realist and intuitionist alike, that a decidable predicate can meaningfully be taken to be applicable to any natural number, however large, he said:

> Now, I feel disinclined, in a way, to say what I'm going to say, and I realise that it is a strong, hard-headed thing to assert.
> You know this famous remark of Wittgenstein about whether a very very large number is prime or composite. Turing said: 'Well, God knows which it is', to which Wittgenstein replied that there is nothing for God to know, because all He knows is the results of our calculations.
> I find I cannot follow this line. I'm more inclined to say that if you have a method which can be used over some small range, as when you determine whether a number is prime or not, and you know how to do this (that is to say: it is a method which is not in itself limited to the range, it is just in practice limited to the range), then I wish to say that you have determined a sense for applying the predicate everywhere. (Pataut 2002, 239)

Because of these three properties, Dummett's inference to A3 is an inference to the best explanation; in view of the third property, it is, more specifically, an inference to the only reasonable explanation, in which some will have greater confidence than in the general case.

While the classification of Dummett's particular inference as an inference to the best explanation is, I think, not problematic, in his writings I have found no methodological comment on the form of this particular inference.[12] Timothy Williamson, however, reports that Dummett in 1979–1980, in a different context, expressed scepticism over inference to the best explanation:

> The difference between us about how to do philosophy is that you think that inference to the best explanation is a legitimate method of argument in philosophy, and I don't. (Williamson 2016, 264)

Williamson takes the idea behind Dummett's view (with which he disagreed) to have been that

> The deepest, most important philosophical questions are about meaning. If someone proposes a philosophical theory to explain some data, there is a prior question as to whether the theory even has a coherent meaning. If not, it offers no genuine explanation. While that question remains unsettled, attempts to apply inference to the best explanation are premature. But once the question of meaning has been settled, the question of explanation will be comparatively trivial. (Williamson 2016, 264)[13]

This will be a very plausible interpretation to any reader of Dummett. In the case at hand, as Dummett presents it, the realist's and the ontological intuitionist's agreement that understanding a decidable statement consists in mastering a decision procedure for it settles enough of the question of meaning to run the argument; and given Dummett's assessment of the subjunctive conditional in A3 as highly natural, he would surely have found the question of explanation, once he would have considered Argument A from that perspective, straightforward. Dummett must have been attracted to the objection to the ontological route he makes because, while fully allowing the coherence of that route, it promises a strong argument against it even without going into genuinely meaning-theoretical considerations; that is indeed the point of the argument. The

---

[12] Wright, who is interested in the alternative use of (what I here call) Argument A as an argument against strict finitism, considers that it does no better than 'diagnose our intuitive opposition' (Wright 1982, 242). I would argue that, as Argument A can be seen as an inference to the best explanation, there is, whatever one ultimately thinks of that form of inference, more to it than Wright allows.

[13] Williamson goes on to pose critical questions about such a view; he advocates a 'broadly abductive methodology' in philosophy himself (Williamson 2016, 264–265 and 268).

application of inference to the best explanation, then, is not an untimely one, for it is hard to see how a case against the ontological route that does not consist in arguing for its logical or conceptual incoherence, nor in a diagnosis of defectiveness on meaning-theoretical grounds, could proceed differently than by making an abduction in favour of an alternative view.[14]

All the same, Dummett would welcome an alternative, meaning-theoretical argument for the conclusion that if we decide a statement now, it had, unbeknownst to us, the corresponding truth value all along. That this is how he saw it I gather from an exchange he had with Prawitz in 1998. Dummett was inspired by Prawitz to consider adopting a notion of constructive mathematical truth according to which truth of a statement consists in our capacity to prove it, whether we ever do this or not, and whether we are aware of having this capacity or not (Dummett 1998a, 1998b; Prawitz 1998b, 1998a; Pataut 2002, 244–245). I will refrain from going into the details of that discussion here; it is wholly internal to the meaning-theoretical approach, and the notion of truth it seeks to develop is, in spite of remaining differences between Prawitz and Dummett, in any case one that contradicts T2. The salient point for my present purpose is that in these reflections, Dummett refers to his own idea that 'it is already determinate how the decision procedure will turn out, because there is no room for any play in the process of applying it' as a 'tacit assumption' (Dummett 1998b, 123), without any suggestion of his prior considerations in 'The philosophical basis of intuitionistic logic'. Yet, Dummett has never disavowed his objection to the ontological argument (see footnote 3 above).

With Argument A in place and accepted, inspection of its structure then leads Dummett to an argument against ontological intuitionism that may be presented as follows:

B1   Argument A is acceptable. (Premiss)
B2   A4 contradicts T2-d and hence T2. (Premiss)
B3   T2 is false. (From B1 and B2)
B4   Argument A is independent of any ontological premiss. (Premiss)
B5   The ontological route to an intuitionistic interpretation of logic is closed. (From B3 and B4)

---

[14]In a similar vein, Williamson writes: 'Dummett held that we can settle general questions of meaning only by establishing a systematic *theory* of meaning. But how can we establish it if not by inference to the best explanation? The uncontested data will not entail any non-trivial systematic theory of meaning.' (Williamson 2016, 264, original emphasis)

The observation in Premiss B4 enables one to relate Argument A to the ontological route; the nature of that relation is then made explicit in conclusion B5. Accordingly, the ontological route to intuitionistic logic is closed, not because T1 would be false, but because T2, which leads to non-classical logic for undecidable statements, is untenable already for decidable ones.

For the debate with the realist, as Dummett construes it, this means that the attempt to make the case that the correct logic for undecidable statements is intuitionistic by first arguing that T2d is correct and then extending this to all statements to arrive at T2 cannot succeed, since it commits one to an unacceptable scepticism. If Dummett is right, then the ontological intuitionist should accept Argument A after all, and see the relation between decidable statements and their truth values as the realist does, as obtaining even before a decision procedure is ever carried out. That view, when extended to undecidable statements, is consistent with the acceptance of either classical logic or intuitionistic logic for them, and therefore cannot force the choice. (As Dummett indicates, even someone accepting that mathematical statements have determinate truth values may still be persuaded to reject classical logic on meaning-theoretical grounds (Dummett 1978a, 246).)

3. A rejoinder

The ontological intuitionist is in a position to reply to Dummett as follows.

Procedures come in two kinds, investigative and generative. A procedure is investigative if that in virtue of which a true proposition stating its outcome is true exists prior to the execution of that procedure. A procedure is generative if the existence of that in virtue of which a true proposition stating its outcome is true is brought about by the execution of that procedure. By T2, the ontological intuitionist holds that all mathematical procedures are generative. For example, assuming that we have no prior knowledge whether a given large number n is prime, then carrying out the procedure of a primality test with n as argument *makes* it prime, if the test is positive, and composite otherwise. The decision procedure is not merely a procedure 'to test out' upon which 'we should find' what is the case (Dummett 1978a, 244); it is a procedure *to bring about*.[15]

---

[15]Although Wright takes decision procedures in mathematics to be investigative, he makes an observation concerning conditionals about non-investigative procedures outside mathematics (Wright 1980, 214–215) that can be transposed to the generative conception of intuitionistic mathematics. If, by hypothesis, we have brought about the truth of some proposition p, we may well thereby have put ourselves in a

Obtaining a truth value by carrying out a decision procedure then is more like baking a cake according to a recipe than like finding a treasure with the guidance of a pirate map.

Although on Brouwer's generative view mathematical objects, their properties, and, thereby, that in virtue of which true mathematical propositions are true, are all brought about by the subject, it cannot bring these about without any constraints. They have to be constructed on the basis of the intuition of time:

> The only possible foundation of mathematics must be sought in this construction under the obligation carefully to watch which constructions intuition allows and which not. (Brouwer 1907, 77, trl. Brouwer 1975, 52)

Obvious concerns have been raised that this view is psychologistic, and that it cannot account for the intersubjectivity of mathematical truth.[16] The potential power of Dummett's objection to ontological intuitionism, however, lies in the fact that it can still be made even if the intuitionist can alleviate such worries. As long as an ontological view validates T2 and Premisses A1 and A2, it makes no difference for Dummett's argument how that view is given further substance. The objection I want to present to Dummett's argument should therefore not turn on that.

The problem with Argument A, once the distinction between the two kinds of procedures is accepted, is that its two premises, A1 and A2, are true for both investigative and generative procedures, but its conclusion is true only for investigative ones, and indeed counts as a paraphrase of the statement that decision procedures are investigative. What has gone wrong along the way is that, while the inference from A1 and A2 to A3 is acceptable for both kinds of procedures, the inference from A3 to A4 is correct for investigative ones only, and therefore begs the question. This diagnosis of Argument A does not require one to hold that the generative conception of mathematical procedures is itself the correct one.

Let us look at this in more detail. There is no discussion contrasting subjunctive and indicative conditionals to be found in Brouwer and Heyting,

---

position to bring about the truth of some proposition q, and knowingly so, without it thereby being correct to say that p implies q. This happens when to bring about p is to make q possible without quite necessitating it. In intuitionistic mathematics, such necessitation is of course part of what is meant by implication. This does not rule out that a method that transforms constructions for p into constructions for q includes making choices at some stages; what will guarantee the necessity of q relative to p in such a case is that the method arrives at a construction for q no matter how these choices turn out. For example, suppose that in a choice sequence a we have so far made only the choices for $\alpha(0)$ and $\alpha(1)$; then we are in a position to make it true that $\alpha(2) = 0$, simply by choosing 0 for the third number in a. But there is no implication that, if in a choice sequence we have made the first two choices, then the third choice will be 0.

[16]See van Atten (2004, chapter 6) and the references there.

but neither of them betrays scepticism about subjunctive conditionals. For example, Brouwer writes,

> Gold voor dit reëele getal de relatie $\rho > 0$, dan zou ρ niet < 0 kunnen zijn (Brouwer 1948, 963)

the subjunctive in which is preserved in the translation by the editor of the *Collected Works*:

> If for this real number ρ the relation $\rho > 0$ were to hold, then $\rho < 0$ would be impossible. (Brouwer 1975, 478)

And Heyting comments on the definition '*l* is the greatest prime such that *l* −2 is also a prime, or *l*=1 if such a number does not exist' that

> if the sequence of twin primes were either finite or not finite, [this] would define an integer. (Heyting 1956, 2)

They are right to do so, because on their generative conception of mathematics, subjunctive conditionals are readily accounted for, as we will now see for the case of A1 and A3.[17] The subjunctive conditional

A1   If the procedure D were carried out, the result would be $P(n)$ or $\neg P(n)$.

can be asserted under exactly the same conditions as the indicative conditional

A1-i   If the procedure D is carried out, the result will be $P(n)$ or $\neg P(n)$.

The reason is that in both cases the relation between the carrying out of the generative procedure and the proposition that its outcome is $P(n)$ or $\neg P(n)$ is that the former brings about that in virtue of which the latter becomes true, by mathematical necessity. Because of this self-sufficiency of the procedure, it makes no difference for the holding of this relation whether the circumstances under which the procedure is carried out are real or imagined, nor where it is situated in time. And A1-i should indeed be acceptable to the ontological intuitionist upon understanding the procedure at all; therefore, A1 is acceptable as well. Of course, one

---

[17] The argument in this paragraph for the acceptability of A1, as well as that of A3 further on in the text, is indebted to the analysis of conditionals in Dudman (1984) and Dudman (1994); the fact that we only have to consider if-clauses about generative mathematical procedures simplifies matters. The questions whether the proximity between Dudman's analysis of conditionals and intuitionistic views on them is even closer than what I make use of here, and where and to what extent the general views on conditionals of Dudman and Dummett disagree, can be left aside here.

would all the same choose to utter A1 instead of A1-i to indicate that one is thinking counterfactually, or to convey that this relation holds independently of exactly what mathematical activity occurred before the moment of utterance (Dudman 1994, 122–123).

Premiss A2 can be readily accepted. But the ontological intuitionist can, even without knowing yet what the result of the procedure will be, furthermore accept the claim that, whenever the same decision procedure were to be carried out with the same input, the result would always be the same. The subjunctive conditional

> If, for a given n, we were to carry out the decision procedure D at different times, then the results would be the same.

can be asserted under exactly the same conditions as the indicative conditional

> If, for a given n, we carry out the decision procedure D at different times, then the results will be the same.

The argument for the conclusion that these can be asserted under exactly the same conditions is similar to that just given for A1 and A1-i: in both of these conditionals, the connection between the generating procedure, a procedure that remains the same across different executions real or imagined, and any proposition about its result when applied to a given n, is the same: if the latter is true, then that in virtue of which it is true has been brought about in the execution of that procedure, at whatever point in time. And once it is given both that the result of carrying out D will always be $P(n)$ or $\neg P(n)$, and that the result will always be the same, then it can be concluded that the result will always be $P(n)$, or that the result will always be $\neg P(n)$. We thus arrive at the indicative version of A3:

A3-i If, for a given n, we carry out the decision procedure D, then we will find that $P(n)$ holds, or, if we carry out the decision procedure D, then we will find that $\neg P(n)$ holds.

In fact, the intuitionist's recognition of the procedure D as a *decision* procedure consists in seeing that A3-i is correct, for otherwise it would not even be known that, given the same argument n, D will not give different answers at different times. In a manuscript of 1951[18] Brouwer

---

[18] Remarks to the same effect are made in print in Brouwer (1955, 114).

remarks about the introduction of the intuitionistic conception of mathematical truth that

> An immediate consequence was that for a mathematical assertion a the two cases of truth and falsehood, formerly exclusively admitted, were replaced by the following three:
>
> (1) α has been proved to be true;
> (2) α has been proved to be absurd;
> (3) α has neither been proved to be true nor to be absurd, nor do we know a finite algorithm leading to the statement either that α is true or that α is absurd.

It is because he accepts A3-i that he can then add this footnote:

> The case that α has neither been proved to be true nor to be absurd, but that we know a finite algorithm leading to the statement either that α is true, or that α is absurd, obviously is reducible to the first and second cases. (Brouwer 1981, 92)[19]

This reducibility is the ontological intuitionist's reason for accepting that the logic of decidable statements is classical,[20] or, better, that for decidable statements classical and intuitionistic logic coincide formally.[21]

From the acceptability of A3-i follows that of A3, because the latter is assertible exactly if the former is; the reasoning here is, again, the same as for A1 and A1-i.[22]

---

[19]See also van Stigt (1990, 450,454).

[20]Wright (1980, 240) states that 'it is likely that the intuitionists' acceptance of classical logic for effectively decidable statements was motivated by, in our terms, a belief in their investigation-independence' (which latter term is shorthand for 'the investigation-independent determinacy in truth-value' (Wright 1980, 206)). That was evidently not the case for Brouwer and Heyting. Be that as it may, Wright clearly thinks that it is better to have that belief:

> Now, certainly, when *R* is a mathematical statement, we should not ordinarily suppose that anything we do is capable of affecting whether or not a correct proof of *R* can be given [...] For this reason we naturally took it that to have verified $P \to R$, *P* hypothesising correct implementation of an effective investigative procedure, is to have verified that we have *access* to a proof of *R*; access, that is, to a construction whose status as a proof of *R* in no sense awaits our actual recognition. (Wright 1980, 213, original emphasis)

This view contradicts T2 directly, whereas in Dummett's objection to ontological intuitionism that is the result of an argument.

[21]When Gödel remarked that his double-negation translation showed that 'the system of intuitionistic arithmetic and number theory is only apparently narrower than the classical one, and in truth contains it, albeit with a somewhat deviant interpretation' (Gödel 1933, 37, trl. Gödel 1986, 295) Heyting replied that 'for the intuitionist, the interpretation is what is essential' (Heyting 1934, 18, trl. mine).

[22]That reasoning also justifies conditionals in which both the condition and that what would be true upon its satisfaction are located in the past (Dudman 1984, 187). This allows a reply to Fregean objections to intuitionism such as these:

> By making mathematics dependent on human thought, even if not on individual acts thereof, we would still ascribe incorrect temporal and modal properties to mathematical truths. As Frege famously argued, all mathematical truths now known were true already before we discovered them, indeed even before the existence of the human species. (Linnebo 2017, 78)

The matter of the acceptability of subjunctive conditionals is brought up by Dummett because he reasons that ontological intuitionism, in rejecting A4, must be rejecting the inference from A1 and A2 to A3. The possibility of a generative reading shows that this need not be the case. The problem with Argument A, however, is that the inference from A3 to A4 is not correct for generative procedures, by definition, and is therefore, unlike premisses A1 and A2, not neutral between the two kinds. As a consequence, the conclusion of Argument A cannot be arrived at on all readings under which its premisses come out true, and Argument A fails. Therefore, premiss B1 is false, and Argument B does not establish its conclusion that the ontological route is closed.

Acknowledgments

Earlier versions were prepared for the 'Michael Dummett Paris Conference', Paris, March 2014, the seminar of the ILLC, Amsterdam, October 2014, and 'Critical Views of Logic', Oslo, August–September 2017. I am grateful to the organisers for their invitations. In Oslo, at the last moment I preferred to present on a different topic, dissatisfied with the manuscript as it was; the organisers kindly permitted me to do so. I thank the audiences for their questions and comments, and Igor Douven and Göran Sundholm for additional conversation. Luca Tranchini kindly helped obtaining some of the literature. Finally, I thank the editors for comments on the penultimate version.

Disclosure statement

No potential conflict of interest was reported by the authors.

References

Auxier, R. E., and L. E. Hahn, eds. 2007. *The Philosophy of Michael Dummett*. Chicago and La Salle, IL: Open Court.

Brouwer, L. E. J. 1907. "Over de grondslagen der wiskunde." PhD thesis, Universiteit van Amsterdam.

The ontological intuitionist can perfectly well give an interpretation of claims such as 'It was true that 3 is odd even before the existence of the human species', by construing it as 'If there would have been a thinking subject at the time who created the number 3 and carried out the parity test, the result would have been that 3 is odd'; and similarly for the claim of the invariance in counterfactual situations of the mathematical truth that 3 is odd. An intuitionistic interpretation of that type would of course be incoherent for counterfactual conditions in which there could be no thinking subject. Such conditions can be meaningful only if a Fregean thought is taken to exist independently not only of any actual act of thinking it, but of any possible act of thinking it. But to separate, conceptually, thoughts from acts of thinking altogether, is, as has been forcefully argued by Dummett in 'Frege's myth of the Third Realm', to succumb to a philosophical myth (Dummett 1986); see also 'More about thoughts' (Dummett 1989, in particular 312–313).

Brouwer, L. E. J. 1948. "Essentieel Negatieve Eigenschappen." *KNAW Proceedings* 51: 963–964.

Brouwer, L. E. J. 1955. "The Effect of Intuitionism on Classical Algebra of Logic." *Proceedings of the Royal Irish Academy* 57: 113–116.

Brouwer, L. E. J. 1975. *Collected Works I. Philosophy and Foundations of Mathematics*, edited by A. Heyting. Amsterdam: North-Holland.

Brouwer, L. E. J. 1981. *Brouwer's Cambridge Lectures on Intuitionism*, edited by D. van Dalen. Cambridge: Cambridge University Press.

Dudman, V. 1984. "Conditional Interpretations of If-sentences." *Australian Journal of Linguistics* 4 (2): 143–204.

Dudman, V. 1994. "On Conditionals." *Journal of Philosophy* 91 (3): 113–128.

Dummett, M. 1959. Truth. In *Proceedings of the Aristotelian Society*, n.s., Vol. 59, 141–162. Page references are to the reprint in Dummett (1978b, 1–24).

Dummett, M. 1963. "The Philosophical Significance of Gödel's Theorem." *Ratio* 5: 140–155. Page references are to the reprint in Dummett (1978b, 186–201).

Dummett, M. 1973. *Frege: Philosophy of Language*. London: Duckworth.

Dummett, M. 1975. The Philosophical Basis of Intuitionistic Logic. In *Logic Colloquium '73*, edited by H. Rose and J. Sheperdson, 5–40. Amsterdam: North-Holland.

Dummett, M. 1976. "What is a Theory of Meaning? (II). In *Truth and Meaning*, edited by G. Evans and J. McDowell, 67–137. Oxford: Clarendon Press. Page references are to the reprint in Dummett (1993b, 34–93).

Dummett, M. 1977. *Elements of Intuitionism*. Oxford: Oxford University Press.

Dummett, M. 1978a. "The Philosophical Basis of Intuitionistic Logic." In Dummett (1978b, 215–247). Reprint of Dummett (1975).

Dummett, M. 1978b. *Truth and Other Enigmas*. London: Duckworth.

Dummett, M. 1980. "Critical Notice. L.E.J. Brouwer: *Collected Works*." *Mind; A Quarterly Review of Psychology and Philosophy* 89: 605–616.

Dummett, M. 1981. *The Interpretation of Frege's Philosophy*. London: Duckworth.

Dummett, M. 1986. "Frege's Myth of the Third Realm." *Untersuchungen zur Logik und zur Methodologie* 3: 24–38. Page references are to the reprint in Dummett (1991b, 248–262).

Dummett, M. 1989. "More About Thoughts." *Notre Dame Journal of Formal Logic* 30 (1): 1–19. Page references are to the reprint in Dummett (1991b, 289–314).

Dummett, M. 1991a. *Frege: Philosophy of Mathematics*. London: Duckworth.

Dummett, M. 1991b. *Frege and Other Philosophers*. Oxford: Oxford University Press.

Dummett, M. 1993a. *Origins of Analytical Philosophy*. London: Duckworth.

Dummett, M. 1993b. *The Seas of Language*. Oxford: Oxford University Press.

Dummett, M. 1994. "Reply to Prawitz." In *The Philosophy of Michael Dummett*, edited by B. McGuinness and G. Oliveri, 292–298. Kluwer.

Dummett, M. 1998a. "Is the Concept of Truth Needed for Semantics?" In Martinez, Rivas, and Vellegas-Forero (1998, 3–22).

Dummett, M. 1998b. "Truth From the Constructive Standpoint." *Theoria* 64: 122–138.

Dummett, M. 2000. *Elements of Intuitionism*. 2nd, rev. ed. Oxford: Clarendon Press.

Dummett, M. 2007a. "Intellectual Autobiography." In Auxier and Hahn (2007, 3–32).

Dummett, M. 2007b. "Reply to Wolfgang Künne." In Auxier and Hahn (2007, 345–350).

Gödel, K. 1933. "Zur Intuitionistischen Arithmetik Und Zahlentheorie." *Ergebnisse eines mathematischen Kolloquiums* 4: 34–38.

Gödel, K. 1986. *Collected Works. I: Publications 1929–1936*, edited by S. Feferman, et al. Oxford: Oxford University Press.

Heyting, A. 1934. *Mathematische Grundlagenforschung, Intuitionismus, Beweistheorie*. Berlin: Springer.

Heyting, A. 1956. *Intuitionism. An Introduction*. Amsterdam: North-Holland.

Heyting, A. 1958. "Intuitionism in Mathematics." In *La philosophie au milieu du vingtième siècle*, edited by R. Klibansky, volume 1, 101–115. La nuova Italia.

Künne, W. 2007. "Two Principles Concerning Truth." In Auxier and Hahn (2007, 315–344).

Linnebo, Ø. 2017. *Philosophy of Mathematics*. Princeton: Princeton University Press.

Martinez, C., U. Rivas, and L. Vellegas-Forero, editors. 1998. *Truth in Perspective: Recent Issues in Logic, Representation and Ontology*. Aldershot: Ashgate.

Pataut, F. 2002. "Truth, Meaning, Modalities, Ethics. (A Second Interview with Michael Dummett)." *Philosophical Investigations* 25: 225–271.

Placek, T. 1999. *Mathematical Intuitionism and Intersubjectivity*. Dordrecht: Kluwer.

Posy, C. 1991. "Mathematics as a Transcendental Science." In *Phenomenology and the Formal Sciences*, edited by T. Seebohm, D. Føllesdal, and J. Mohanty, 107–131. Kluwer.

Prawitz, D. 1998a. "Comments on the Papers." *Theoria* 64: 283–337.

Prawitz, D. 1998b. "Truth from a Constructive Perspective." In Martinez, Rivas, and Vellegas-Forero (1998, 23–35).

Rose, H., and J. Sheperdson, editors. 1975. Logic Colloquium'73. Amsterdam: North-Holland.

Sundholm, G. 1994. "Existence, Proof and Truth-making: a Perspective on the Intuitionistic Conception of Truth." *Topoi* 13 (2): 117–126.

van Atten, M. 2004. *On Brouwer*. Belmont: Wadsworth.

van Stigt, W. 1990. *Brouwer's Intuitionism*. Amsterdam: North-Holland.

Williamson, T. 2016. "Abductive Philosophy." *The Philosophical Forum* 47 (3–4): 263–280.

Wright, C. 1980. *Wittgenstein on the Foundations of Mathematics*. London: Duckworth.

Wright, C. 1982. "Strict Finitism." *Synthese* 51 (2): 203–282.

The entanglement of logic and set theory, constructively

Laura Crosilla

ABSTRACT
Theories of sets such as Zermelo Fraenkel set theory are usually presented as the combination of two distinct kinds of principles: logical and set-theoretic principles. The set-theoretic principles are imposed 'on top' of first-order logic. This is in agreement with a traditional view of logic as universally applicable and topic neutral. Such a view of logic has been rejected by the intuitionists, on the ground that quantification over infinite domains requires the use of intuitionistic rather than classical logic. In the following, I consider *constructive* set theories, which use intuitionistic rather than classical logic, and argue that they manifest a distinctive *interdependence* or an *entanglement* between sets and logic. In fact, Martin-Löf type theory identifies fundamental logical and set-theoretic notions. Remarkably, one of the motivations for this identification is the thought that classical quantification over infinite domains is problematic, while intuitionistic quantification is not. The approach to quantification adopted in Martin-Löf's type theory is subtly interconnected with its predicativity. I conclude by recalling key aspects of an approach to predicativity inspired by Poincaré, which focuses on the issue of correct quantification over infinite domains and relate it back to Martin-Löf type theory.

1. Introduction

Standard approaches to set theory such as Zermelo Fraenkel set theory (ZF) take classical first-order logic as granted and specify what counts as a set by means of a collection of axioms imposed 'on top' of the classical predicate calculus.[1] In fact, ZF set theory is often presented as if it were obtained by combining two wholly separable layers: the logical and the set-theoretic axioms. This view of set theory as composed of two layers,

---

[1]The theory ZF is composed of axioms such as, for example, the power set axiom, and schemata such as, for example, the replacement schema. In the following, I use 'principle' or 'axiom' to refer to either an axiom or a schema.

first the logic and then the specific set-theoretic axioms, is in agreement with traditional accounts of logic, according to which distinctive characteristics of logic are its *generality* and *universal applicability*. Logic, according to this view, is *topic neutral*: the particular domain under consideration has no say on the logical principles appealed to when reasoning about its objects.

This view of logic as applicable irrespective of the relevant domain of discourse has not gone unchallenged. Traditionally, intuitionists have taken a critical perspective on the universal applicability of classical logic. A frequent objection is that classical logic is not the correct logic to employ when reasoning about *infinite domains*: for those domains, we need to use *intuitionistic logic*. This criticism may be seen as underpinning the thought that different domains of investigation may require different logics. The intuitionist will typically claim that given the pervasive role of infinity in mathematics, *in mathematics* we ought to use intuitionistic logic.

My starting point in this article is the mathematical practice. Under the stimulus of computer application, contemporary mathematics has seen the flourishing of *constructive* mathematics, which employs intuitionistic rather than classical logic.[2] In the following, I review prominent constructive approaches to the concept of set and argue that they question the above image of the relation between sets and logic, rather suggesting a distinctive *interdependence* or an *entanglement* between sets and logic. I consider two kinds of theories: constructive variants of ZF, such as CZF, and Martin-Löf Type Theory, MLTT.[3] Constructive set theories of the ZF family question the very possibility of sharply distinguishing set-theoretical and logical principles. In MLTT one further *identifies* fundamental logical and set-theoretic notions: *proposition* and *set*.[4] Crucially, the identification of propositions and sets in MLTT is deeply interconnected with the computational vocation of this theory, which aims at offering not only a foundational system for constructive mathematics but also a theory for (computer) program construction. As argued below, MLTT's identification of propositions with sets, which goes under the name of Curry-Howard

---

[2]There are a number of variants of mathematics carried out using intuitionistic logic. The best-known constructive practice, and the one I shall refer to in the following, is the mathematics initiated by Bishop (1967). See also Beeson (1985); Bishop and Bridges (1985); Bridges and Richman (1987); Troelstra and van Dalen (1988). For an introduction and further bibliographic references, see Bridges and Palmgren (2013).

[3]See Martin-Löf (1975); Myhill (1975); Aczel (1978); Martin-Löf (1982); Martin-Löf (1984); Aczel and Rathjen (2008).

[4]I discuss MLTT's notions of proposition and set in Section 4.

isomorphism, cannot be explained solely with Martin-Löf's desire to develop a computational system for constructive mathematics. A further motive for this identification of logical and set-theoretic notions is the desire to fully embrace the intuitionistic meaning of the logical constants.

It is perhaps surprising that in the opening of the book 'Intuitionistic type theory' (Martin-Löf 1984), the author explains the choice to develop an *intuitionistic* type theory with a perceived difficulty with classical quantification over *infinite domains* and relates this back to the early twentieth-century debate on *predicativity*. Predicativity made its appearance in an exchange between Poincaré and Russell, prompted by the discovery of the set-theoretic paradoxes.[5] According to one rendering of this notion, a definition is impredicative if it defines an entity by reference to (e.g. generalisation over) a totality to which the entity itself belongs, and is predicative otherwise. Martin-Löf's type theory is said to be predicative as higher order quantification over all propositions, all relations etc. is unavailable within this theory. There is, in fact, a deep connection between MLTT's predicativity and the Curry-Howard isormorphism, as further discussed below. There is therefore a deep connection between MLTT underlying intuitionistic logic and this theory's predicativity.

A strong correlation between predicativity and the issue of legitimate quantification over infinite domains is a theme which appears prominently in the reflection on predicativity by Poincaré (1909, 1912).[6] Poincaré saw a strong nexus between the insurgence of the paradoxes, impredicativity and what he considered as incorrect treatment of infinite domains. Adopting a traditional view of infinity as potential rather than actual, Poincaré thought that paradoxes like Russell's arise because we treat as actual or completed potentially infinite domains. He proposed predicative restrictions as a way of ensuring a paradox-free treatment of infinite domains. Poincaré's discussion on predicativity offers a possible way of explaining the perceived difficulties with classical quantification over infinite domains; the identification of propositions with sets in Martin-Löf's type theory offers a way out from these difficulties.

The article is organised as follows. In Section 2, I review the standard approach to set theory exemplified by ZF set-theory. I then review significant aspects of constructive Zermelo Fraenkel set theory, CZF, and compare this system with ZF set theory (Section 3). The principal claim is that constructive variants of ZF place under strain the very separation

---

[5]See e.g. Poincaré (1905, 1906, 1909, 1912); Russell (1906a, 1906b, 1908). See also Feferman (2005); Crosilla (2017).
[6]This is also a prominent theme in Weyl (1918).

between set-theoretic and logical principles. Starting from Section 4, I discuss Martin-Löf's type theory and consider the relation between the logical and set-theoretic notions of proposition and set. In Section 6, I discuss the relation between this identification and the computational vocation of constructive type theory. In Section 7, I review Martin-Löf's discussion of quantification over infinite domains. Finally, in Section 8, I highlight the main traits of an analysis of predicative definitions by the late Poincaré and briefly discuss MLTT's treatment of quantification.

2. Zermelo Fraenkel set theory and its family

In introducing Zermelo Fraenkel set theory, we usually present it as if it were the combination of two wholly distinct layers: the logical and the set-theoretic axioms. The first layer takes the form of the first-order predicate calculus, detailing the behaviour of the logical constants and the quantifiers: it specifies legitimate ways of reasoning. The set-theoretic axioms, instead, determine the subject matter of the theory and state which sets the theory is about. More precisely, the set-theoretic axioms may be divided into three kinds of principles: some introduce primitive sets (e.g. the empty set and the set of natural numbers), others specify operations for forming new sets from already given ones (e.g. union, power set, etc.) and others set out the identity criteria for sets (extensionality), as well as properties of the universe of sets (foundation). The two components of ZF, logical and set-theoretic axioms, are typically perceived as separate: the set-theoretic axioms are imposed 'on top' of the first-order predicate calculus. In fact, standard presentations of ZF focus primarily on the concept of set: sets have a foundational and unifying role to play, as, arguably, any mathematical object can be defined in terms of them.[7] While classical logic is presupposed and given for granted, we need a formal logical calculus to remove possible ambiguities, especially in the formulation of the separation schema.[8] Given this way of presenting set theory, it is tempting to think that the set-theoretic principles we use to codify ZF's concept of set are clearly distinguishable from the background logical ones: their role is to systematize the concept of set, rather than clarify how to reason about sets. This may be further substantiated by observing that a number of variants of ZF set theory have been considered in the literature. These are obtained by taking a different combination of

---

[7]See e.g. Suppes (1960); Hrbacek and Jech (1999).
[8]The separation schema allows for the definition of subsets of a given set, A, by collecting together all the elements of A that satisfy a given property, expressed by a formula in the language of set theory.

set-theoretic axioms on the basis of the *same* first-order logic. As a way of example, a remarkable subsystem of ZF, Kripke Platek set theory, features prominently in generalised recursion theory, while extensions of ZF by large cardinal axioms are at the forefront of research in set theory. The set-theoretic axioms may be read as specifying which kind of sets one is concerned with (e.g. ZF sets, admissible sets, large cardinals, etc.). The very possibility of defining *different* set theories on the basis of the *same* logical system may be taken to suggest some form of independence of the logical from the set-theoretic principles.

This way of presenting set theory is in agreement with a traditional conception of logic as universally applicable, wholly general and topic neutral. According to this view, the laws of logic govern the forms of thought irrespective of the particular domain of discourse we are concerned with, they are universally applicable and do not belong to the sphere of some special science.[9] In particular, the logical laws hold of any domain, irrespective of the nature of its elements and, especially, its cardinality.

Very different is the view expressed by the intuitionist. For the latter, there is a substantial agreement between intuitionistic and classical forms of mathematics when we consider finite sets; however, a profound divergence appears when we consider infinite sets.[10] From an intuitionistic perspective, classical concepts relating to infinite or transfinite sets become meaningless and may even give rise to inconsistency. For this reason, mathematics requires the adoption of intuitionistic rather than classical logic. As further discussed below, a stark opposition between the finitary and the infinitary cases is also at the heart of Poincaré analysis of impredicativity (1909, 1912). This criticism of classical logic is clearly conveyed by Weyl, who also draws a correlation between the use of *classical logic* on infinite domains and the paradoxes of set theory:

> [...] classical logic was abstracted from the mathematics of finite sets and their subsets. ... Forgetful of this limited origin, one afterwards mistook that logic for something above and prior to all mathematics, and finally applied it, without justification, to the mathematics of infinite sets. This is the fall and original sin of set theory, for which it is justly punished by the antinomies. It is not that such contradictions showed up that is surprising, but that they showed up at such a late stage of the game. (Weyl 1946, 9-10)

More recently, Dummett (1991, 1993) has also argued that quantification over infinite domains requires not classical but intuitionistic logic. As

---

[9]See e.g. Frege (1953).
[10]See, for example, Brouwer (1912).

further discussed below, the desire to make sense of quantification over infinite domains figures also in the constructive approach to the concept of set put forth by Martin-Löf.

3. Constructive ZF

Recent decades have seen the rise of variants of ZF which use intuitionistic rather than classical logic, such as, for example, Intuitionistic Zermelo Fraenkel set theory (IZF) and Constructive Zermelo Fraenkel set theory (CZF).[11] These theories have been proposed as ways of formalising, in the familiar ZF language, the concept of set that is implicit in the work of constructive mathematicians. Prima facie, here too one has two layers: the logic, intuitionistic in this case, and the set-theoretic axioms. Formulating variants of ZF based on intuitionistic logic, however, requires a careful reconsideration of the ZF axioms, to avoid to accidentally slide back into classical logic. Particular care is required, for example, in formulating the *foundation* axiom, which, in its standard form, states that every non-empty set has a least element with respect to the membership relation. In fact, on the basis of the other principles of ZF and intuitionistic logic, foundation implies the principle of excluded middle.[12] Similarly, adding the full *axiom of choice* to an intuitionistic variant of ZF gives back the principle of excluded middle, which becomes now provable from the axiom of choice (on the basis of the axioms of ZF and intuitionistic logic). Therefore, the axiom of choice needs to be omitted if we aim at formulating an *intuitionistic* variant of ZF set theory.[13]

The case of constructive versions of ZF is suggestive of a form of interaction between the logical and the set-theoretic levels: if we are not careful in formulating our set-theoretic principles, we might engender a sudden change in logic, from intuitionistic to classical logic. Let us consider again the case of foundation. The foundation axiom states that the

---

[11]Thy system IZF was introduced in Friedman (1973). It is an impredicative intuitionistic variant of ZF set theory. For bibliographic references on CZF, see footnote 3.

[12]To be more precise, the axiom of foundation implies the principle of excluded middle on the basis of intuitionistic logic and the other axioms of a system, such as IZF, with the unrestricted separation schema. In the case of systems such as CZF, which only admit separation for bounded formulas, one obtains a bounded form of excluded middle: A ∨ ~A for every bounded formula A. Note that a formula, A, is bounded if every occurrence of quantifiers in A is of the form $\exists x \in D$ or $\forall x \in D$, with D a set. See Aczel and Rathjen (2008); Crosilla (2015).

[13]One may add weaker choice principles such as countable and dependent choice to constructive variants of ZF without giving rise to a classical set theory. It is interesting to observe that while the full axiom of choice cannot be added to intuitionistic systems in the style of ZF, as it gives rise to a classical theory, the situation is different in the case of MLTT. Here a statement expressing the axiom of choice (i.e. an intensional form of the axiom of choice) is in fact provable in virtue of the Curry-Howard isomorphism. See Martin-Löf (2006) for discussion.

membership relation is well-founded, and thus describes a property of sets: well-foundedness. As a consequence, non-well-founded sets, such as sets which are members of themselves, are ruled out from ZF set theory. Foundation would therefore seem to squarely fit within the set-theoretic component of ZF set theory. For this reason, it is surprising that its statement (on the basis of intuitionistic logic and the other axioms of ZF) already implies a logical principle, the principle of excluded middle. In fact, constructively unacceptable instances of the principle of excluded middle can be derived from foundation on the basis of a weak subsystem of CZF, prompting the thought that the axiom of foundation, in its usual formulation, is 'inherently classical'.

One may be tempted to dismiss this case as constituting insufficient proof of a genuine interplay between sets and logic since we can replace foundation with a schema, *set induction*, which is classically equivalent to foundation but does not give rise to the above difficulties in an intuitionistic context. Set induction legitimates reasoning by induction on the membership relation, but it has the advantage of not enforcing classical logic. The thought is that adequate care in formulating our axioms and schemata may suffice to reinstate, even in an intuitionistic context, the separation between logical and set-theoretic principles that we typically envisage in ZF and its classical variants. However, a different situation arises in the case of the axiom of choice, since no equivalent formulation has been proposed which remains neutral with respect to the logic. Only substantial weakening of the axiom of choice (such as countable or dependent choice) may be added to constructive versions of ZF without enforcing the shift to classical logic.[14]

Since the axiom of choice is usually considered set-theoretic rather than logical, it becomes difficult to draw a clear demarcation between set-theoretic and logical principles within constructive theories of the ZF family. This prompts the thought that the set-theoretic component of ZF already embodies a thoroughly *classical* concept of set. These observations are indicative that the very distinction between sets and logic may be more difficult to formulate than it is often thought, and should not be taken for granted. There seems to be, on the contrary, a deep interaction between logic and our concept of set. This is after all unsurprising, as an interaction must arise between the two since sets act as domains of quantification and the concept of set plays a fundamental role in the specification of the semantics of the quantifiers. In fact, it is in relation to the

---

[14]See e.g. Aczel and Rathjen (2008); Crosilla (2015).

role of sets as domains of quantification that Martin-Löf type theory introduces a characteristic identification of crucial logical and set-theoretic notions, as further discussed below.

4. Martin-Löf type theory and the Curry-Howard correspondence

In Martin-Löf Type Theory, we witness a profound interaction between logic and sets (Martin-Löf 1975, 1984). Martin-Löf Type Theory is often considered the *canonical* constructive theory, playing a similar role for constructive mathematics as ZFC does for classical set theory. At the heart of MLTT is an *identification* of two kinds of entities: *propositions* and *sets*. This identification is the Curry-Howard isomorphism. In fact, a weakening of it, often termed Curry-Howard (or formulas-as-types) correspondence is typical of a number of intuitionistic type theories. I discuss MLTT's notions of propositions and sets in Section 5 below. In the following, I first describe the Curry-Howard correspondence in a particularly simple case.

The Curry-Howard correspondence exploits a structural similarity between intuitionistic logic on the one side and constructive type theories on the other side. The correspondence holds already for the implicational fragment of intuitionistic logic, which is seen to correspond to a formalism known as the *typed lambda calculus*.[15] In this particularly simple case, the Curry-Howard correspondence is motivated by the observation that the behaviour of the intuitionistic implication is analogous to the behaviour of a function or program that given inputs (of some type) produces outputs (of some type).[16] The crucial thought is that the domain and range of such a function are types whose elements are (intuitionistic) proofs of formulas. More precisely, the formulas *A* and *B* are seen as corresponding to the types of their proofs, which we can denote by $\underline{A}$ and $\underline{B}$ (respectively). An intuitionistic proof of an implication, $A \to B$, behaves as a *function* which transforms a proof of the formula *A*, into a proof of the formula *B*. That is, an intuitionistic proof of $A \to B$ may also be seen as corresponding to a function which takes as input an element of the type $\underline{A}$, i.e. a proof of the formula *A*, and gives as output an element of

---

[15]See Church (1940); Barendregt (1981).
[16]It is important to remark that the notion of type utilised in constructive type theories does not coincide with the familiar notion of set from classical ZF, as further clarified in Section 4. Furthermore, the notion of function is different from that of ZF. In ZF, a function is a graph, i.e. a set of ordered pairs. In constructive type theories one has a primitive notion of function, and functions may also be thought of as algorithms or procedures that given inputs of some type produce outputs of some type.

the type $\underline{B}$, i.e. a proof of the formula B.[17] Therefore, given the correspondence of A and B with the types of their proofs, the formula $A \to B$ can be seen as corresponding to the type collecting the proofs of this implication. The latter is the collection of all functions with domain $\underline{A}$ and range $\underline{B}$, and is also written $\underline{A} \to \underline{B}$.

The significance of this structural correspondence between formulas (and their proofs) and types (and their elements) lies in that the typed lambda calculus is the core of prominent functional programming languages. It is also at the heart of constructive type theories such as Martin-Löf type theory and the calculus of constructions. These can be read simultaneously as foundational systems for constructive mathematics and as high-level programming languages.[18] A detailed description of MLTT is beyond the aims of this note, but a brief reminder of its main traits is necessary to clarify the identification of propositions with sets.[19]

5. Sets, propositions and their identification

Sets in MLTT are particular kinds of types, those types which function as *domains of quantification*: (typed) collections of 'objects' we can meaningfully quantify over. Martin-Löf traces the origins of his concept of set back to Russell's concept of type as *range of significance of a propositional function* (1984, 22). As further clarified below, Martin-Löf substantially departs from Russell, since MLTT is an *intuitionistic* rather than a *classical* type theory. A characteristic of type theories such as MLTT is that they are not defined by first setting out intuitionistic predicate logic, and then introducing appropriate set-theoretic principles 'on top' of the logical principles. Sets and logic are introduced *simultaneously* through **rules** which define both the logical and the relevant set-theoretic notions. In particular, sets in MLTT are introduced by rules which specify what counts as an element of a set and when two such elements are equal. In virtue of the Curry-Howard isomorphism, the very same rules can also be read as specifying what is a proposition in terms of what counts as a proof of that proposition.

Similarly as in the case of ZF, we may distinguish three groups of rules: those which introduce new sets (e.g. the set of natural numbers), rules

---

[17] One may also say that in constructive type theories we are interested in *proofs* of formulas as ways of witnessing their truth. For example, a proof of an implication, $A \to B$, may be seen as an effective way of transforming a witness of the truth of A (i.e. a proof of A) into a witness of the truth of B (i.e. a proof of B). See Sundholm (1994) for further discussion.

[18] The calculus of constructions is the formal system that underlies the core of the proof assistant Coq (Coquand 1985; Coquand and Huet 1986).

[19] See e.g. Martin-Löf (1984); Nordström, Petersson, and Smith (1990).

which combine previously given sets (e.g. the Cartesian product) and rules which determine relevant identity criteria between sets (equality rules). Within the formalism itself, one has four kinds of rules for each new set or set constructor; for example, the natural number set is defined by four rules and so is the Cartesian product. The rules are *formation, introduction, elimination* and *equality*. The *formation rules* declare which sets there are. For example, they state that there is a set N of natural numbers, or that if A and B are sets, then so are A × B and A→B. One has also rules for the so-called generalised Cartesian product, Π(A,B), and the generalised disjoint union, Σ(A,B), which endow MLTT with particular expressive strength.[20]

Key rules are the *introduction* and *elimination* rules. The purpose of the introduction rules is to declare what is a set, by specifying its *(canonical) elements*.[21] For example, in the case of the natural numbers, N, a canonical element is either 0 or it is of the form suc(n), for n canonical element of N (where 'suc' stands for the successor operation). In the case of the set A × B (the Cartesian product of the sets A and B), a canonical element is a pair, whose first component is an element of A and whose second component is an element of B. The elimination rules explain how to use the elements of a set, that is, they explain how to define functions on the set defined by the introduction rules. Given the identification between propositions and sets, it is unsurprising that most rules governing set-introduction and set-elimination are in fact similar to the natural deduction rules for the introduction and elimination of the connectives and quantifiers in *intuitionistic* predicate logic. In fact, as further discussed below, MLTT's rules are more informative than the corresponding natural deduction rules, as they also carry information on proofs. The equality rules, finally, furnish the relevant identity criteria for sets, by showing how a function defined through the elimination rule acts on the canonical elements of the set which were specified in the introduction rules.

Propositions play a similar role within Martin-Löf's system as formulas in ZF: we may combine propositions by means of the logical operations to give rise to more complex propositions. For example, propositions A

---

[20] The rules for the generalised Cartesian product and the generalised disjoint union make use of the notion of family of sets over a given set: for example, for set A, B(x) is a family of sets over A if B(x) is a set for each element x of A. Under the Curry-Howard correspondence, the generalised Cartesian product, Π(A,B), and the disjoint union, Σ(A,B), of a family of sets correspond to the universal and existential quantifiers (respectively). These constructors therefore allow us to extend the Curry-Howard correspondence from propositional to full predicate logic.

[21] Canonical elements are prototypical or standard elements of a set. Sets typically contain also non-canonical elements, which can be brought to canonical form through application of the elimination rules.

and B can be combined to give rise to the proposition A ∧ B. Propositions may, therefore, be understood as the key logical notion within type theory. A substantial difference with formulas of ZF, however, is that propositions are *mathematical objects*, that is, in MLTT we reason about them within the 'object language', rather than in the 'metalanguage'.[22] In MLTT one identifies propositions and sets. As a consequence, the rules for sets may also be read as rules for propositions. For instance, the introduction and elimination rules for A × B may also be read as the natural deduction rules for conjunction introduction and elimination (respectively).[23] The crucial difference with the usual natural deduction rules for intuitionistic logic, however, is that rules in MLTT are more informative. For simplicity, let us consider the introduction rules only. If we read an introduction rule as introducing a set, then, as we have seen above, the rule specifies the (canonical) elements of that set (and when two such elements are equal). In virtue of the Curry-Howard correspondence between propositions and sets, we may read the same set-introduction rule as introducing a proposition, whose proofs belong to the set. The rule now is read as specifying (canonical) proofs of the proposition.[24] For instance, if we read A × B not as a set but as a proposition, i.e. as the conjunction A ∧ B, its introduction rules first of all state that if A and B are true then so is A ∧ B. More importantly, the rules also specify (canonical) proofs of this proposition: pairs whose components are a proof of A and a proof of B. This additional information (compared with standard natural deduction rules) is at the heart of MLTT's computational interpretation.[25] Clearly, if we are only interested in recovering the natural deduction rules, we can ignore or 'forget' the additional information.[26] For instance, to ascertain the truth of the proposition A ∧ B, it suffices to know that there is a proof of it (or that the set of its proofs is non-empty): we can ignore the additional information detailing the specific form of its proofs.

In Martin-Löf (1975, 1984), the Curry-Howard isomorphism is presented by observing that propositions and sets follow the same pattern of introduction and elimination rules and, on this ground, may be identified. The idea further developed in Martin-Löf (1982) is that the correspondence

---

[22]More precisely, in type theory one distinguishes between *propositions* and *judgements*. The first ones are objects within the theory; we combine them by means of the logical operations. The premises and conclusion of a logical inference are instead judgements (Martin-Löf 1984, 3). For example, if A and B are propositions, the judgment 'A ∨ B is true' asserts the truth of the proposition 'A ∨ B'.

[23]Note that the Cartesian product of two sets, A and B, is obtained by applying the rules for the Σ type constructor, which corresponds to the existential quantifier.

[24]The notion of canonical proof may be explained in terms of direct proof (Martin-Löf 1987).

[25]See, for example, the discussion in Martin-Löf (1982).

[26]See also Sambin and Valentini (1998).

makes the computational or algorithmic content of constructive proofs fully explicit so that we can use MLTT as a very general programming language.[27] This is discussed next.

6. Computational content and predicativity

The Curry-Howard correspondence is a fundamental component of a number of intuitionistic type theories, including the type lambda calculus mentioned above. Most discussions of the Curry-Howard correspondence focus on a weaker claim compared with MLTT's identification of propositions and sets:

(1) every proposition gives rise to a set, the set of its proofs.

This turns out to suffice for a computational understanding of constructive type theories. As a way of clarifying the significance of (1), we can read a proposition as setting out a task or a problem.[28] For example, the proposition stating that every even natural number is the double of some natural number may be seen as setting out the following task: find the half of any even natural number. If we give a *constructive* proof of this proposition, we obtain a step-by-step specification of a procedure that enables us to compute the half of any even natural number. In other terms, a constructive proof of this proposition describes an algorithm which calculates a solution to the problem set by that proposition. By (1), to a proposition corresponds a set, the set of its proofs. This means that a proof of the above mathematical statement on even natural numbers gives rise to a function within type theory. A function in constructive type theory is, after all, an algorithm. In this case, an algorithm that takes even natural numbers as inputs and produces natural numbers as output, their half. In other terms, by working in a constructive type theory that satisfies (1), we produce proofs of mathematical statements and, crucially, these proofs correspond to algorithms that can be used to obtain real computer programs.

A distinctive characteristic of Martin-Löf's type theory is that it strengthens statement (1) to a full *identification* of propositions and sets. Not only each proposition gives rise to a set, but each proposition *is* a set:

---

[27]This thought has been brought substantially forward in recent years through the development of proof assistants such as Nuprl, Agda and Coq.
[28]See Martin-Löf (1984).

[...] there appears to be no fundamental difference between propositions and types. Rather the difference is one of point of view: in the case of a proposition, we are not so much interested in what its proofs are as in whether it has a proof, that is, whether it is true or false, whereas in the case of a type, we are of course interested in what its objects are and not only in whether it is empty or nonempty. (1975, 77)[29]

The identification of propositions and sets that is characteristic of MLTT has far-reaching consequences for the ensuing concept of set that this theory codifies. In fact, combining such an identification with a form of impredicativity gives rise to Girard's paradox.[30] Impredicativity here takes the form of quantification over propositions. More precisely, we may consider an impredicative extension of MLTT obtained by postulating that the collection of all propositions, that we call 'Prop', is itself a set. This gives rise to inconsistency for the following reason: if Prop is a set, i.e. a domain of quantification, and sets are identified with propositions, then Prop acts as the set of all sets or as a universal domain of quantification. A set of all sets figured in a first inconsistent version of MLTT, quickly rectified after the discovery of Girard's paradox. One may then read Girard's paradox as proving the incompatibility of the above form of impredicativity with the identification of propositions and sets. Given such an incompatibility, there are therefore two options: (a) to hold to the identification of propositions and sets and renounce to impredicative quantification, or (b) to renounce to the above identification and allow instead impredicativity.[31] Martin-Löf type theory opts for (a), as discussed in the next section. Option (b) has been pursued with the calculus of constructions, which is an impredicative constructive type theory. To avoid Girard's paradox, the calculus of constructions distinguishes between propositions and sets, which are introduced by separate sets of rules. However, to preserve the computational character of constructive proofs, the calculus of constructions does comply with the Curry-Howard correspondence in the weaker form of (1): instead of the principle that every proposition *is* a set, we simply assume that *for every proposition, there is a corresponding set*, i.e. the set of its proofs. Since propositions and sets are now distinct, this seems sufficient to avoid Girard's paradox while preserving the computational advantages of the Curry-Howard

---

[29]Note that above I have followed (Martin-Löf 1984) and called 'set' what in this quotation from the earlier article Martin-Löf calls 'type'.
[30]See Girard (1972). See also Coquand (1986); Jacobs (1989) for discussion.
[31]See Coquand (1989) for an analysis.

correspondence. In addition, one also gains more expressive power compared with MLTT, due to impredicative quantification.

7. Martin-Löf type theory, infinite domains and predicativity

In MLTT, one takes the narrower route, which avoids impredicativity. Two main components seem to be in play in determining this choice: (i) the thought that a thorough intuitionistic approach requires compliance with the Curry-Howard isomorphism and (ii) the worry often expressed by the intuitionist that classical quantification over infinite domains lacks meaning. As to the first point, Martin-Löf (2008) clarifies that the Curry-Howard isomorphism:

> [...] was to me from the beginning the natural completion of the Brouwer-Heyting-Kolmogorov interpretation of intuitionistic first-order predicate logic, just drawing the full consequences, so to say, of the Brouwer-Heyting-Kolmogorov interpretation.

The thought is that full adherence to intuitionistic logic requires the identification of propositions and sets.[32] By Girard's paradox, compliance with this strong form of Curry-Howard correspondence requires also compliance with predicativity. In fact, the two issues, (i) and (ii), are intimately connected. Martin-Löf (2008) writes:

> Now the paradox that Girard discovered in 1971 (Girard 1971), in the first version of constructive type theory, then called intuitionistic type theory, showed that Curry-Howard is incompatible with impredicativity [...]. So one of them has to go: either Curry-Howard has to go or there is some problem with impredicativity, with which there had been problems from the very beginning: when the notion itself was introduced by Russell in 1906 (Russell 1906), it was precisely because it was a problematic notion.

In his 1984 book (page 1), Martin-Löf also refers back to the origins of predicativity and mentions the perceived difficulty with classical quantification over infinite domains. The author begins by recalling the difficulties caused by the introduction of the axiom of reducibility in *Principia Mathematica* and explains that this left us with the simple theory of types, whose official justification rests on the interpretation of propositions as truth values.

> The laws of the classical propositional logic are then clearly valid, and so are the quantifier laws, as long as quantification is restricted to finite domains. However,

---

[32] A detailed discussion of this point would require an analysis of the most distinctively philosophical component of MLTT, its meaning explanations. Due to space constraint, this will need to be postponed to another occasion. See also Sundholm (1986).

it does not seem possible to make sense of quantification over infinite domains, like the domain of natural numbers, on this interpretation of the notions of proposition and propositional function. For this reason, among others, what we develop here is an intuitionistic theory of types, which is also predicative (or ramified).

Like the intuitionist, Martin-Löf claims that there are problems with classical quantification over infinite domains, and proposes the shift to intuitionistic logic as a solution. At page 11 of Martin-Löf (1984), we read:

> Because of the difficulties of justifying the rules for forming propositions by means of quantification over infinite domains, when a proposition is understood as a truth value, this explanation is rejected by the intuitionists and replaced by saying that a proposition is defined by laying down what counts as a proof of that proposition, and that a proposition is true if it has a proof, that is, if a proof of it can be given.

This then brings us back to the Curry-Howard isomorphism (Martin-Löf 1984, 13):

> If we take seriously the idea that a proposition is defined by laying down how its canonical proofs are formed [...] and accept that a set is defined by prescribing how its canonical elements are formed, then it is clear that it would only lead to unnecessary duplication to keep the notions of proposition and set (and the associated notions of proof of a proposition and element of a set) apart. Instead, we simply identify them, that is, treat them as one and the same notion. This is the formulae-as-types (propositions-as-sets) interpretation on which intuitionistic type theory is based.[33]

The thought that we need to change logic to make sense of quantification over infinite domains clearly disagrees with the traditional view of classical logic as universally applicable and topic neutral which was discussed in Section 2. In fact, the infinitary character of mathematical domains has here a say on which logic we employ in our mathematical practice. A remarkable aspect of Martin-Löf's approach to the issue of quantification over infinite domains is that, in conjunction with the Curry-Howard isomoprhism, it enforces not only a shift to intuitionistic logic but also the theory's compliance with predicativity.

The complaint that classical quantification is unsuitable in the case of infinite domains is a prominent element of the intuitionistic criticism of classical logic. Although Martin-Löf refers to Russell, concerns for quantification over infinite domains are at the heart of the reflection by the other

---

[33] A similar view is presented in Martin-Löf (1975), 76–77, where the author writes: 'it will not be necessary to introduce the notion of proposition as a separate notion because we can represent each proposition by a certain type, namely, the type of proofs of that proposition'.

key historical figure of predicativity. Martin-Löf does not expand on what makes classical quantification over infinite domains problematic. In the following, I wish to briefly recall the main traits of Poincaré's reflection on predicativity (Poincaré 1909, 1912).[34] I will extract a few significant points from Poincaré's analysis which may help us clarify the perceived difficulty with classical quantification over infinite domains and the concomitant requirement of compliance with predicativity. I will then go back to the case of Martin-Löf's type theory.

8. Poincaré's logic of infinity

The focus of Poincaré (1909, 1912) is the question of whether standard ways of reasoning can be applied to infinite collections. The discussion centres on problematic impredicative definitions, which typically specify an entity, say X, by reference to *all* the elements of a collection to which X belongs. Poincaré's worry is that in case the collection we use to define X is infinite, it may have an element of instability: it seems that by reference to it we may give rise to a 'new' element of it, namely X.

Let us clarify this point with an example. Suppose we define an entity, X, by postulating a relation R between this entity and all the elements of an infinite collection defined by a general condition G. Furthermore, suppose that X itself satisfies condition G. Poincaré claims that X's definition is (in general) illegitimate. The thought is that since X is defined by postulating a relation, R, with *all* of the Gs, X's definition would require us to first fix the extension of G, i.e. the collection of all the objects that satisfy G. However, as X turns out to be an element of the extension of G, it is as if X's definition would generate a new element, X, of G's extension. Poincaré also expresses this by claiming that the definition of X may *modify* G's extension and *disorder G* itself.

Poincaré's talk of sets being 'modified' and 'disordered' by the introduction of 'new' elements is clearly problematic from a contemporary classical point of view. It seems to introduce a dynamic component to the concept of set that is extraneous to the classical conception. More should be said on this issue. In the following, I focus on two main components that can be extracted from Poincaré's discussion and may offer a way of making sense of a predicativist perceived difficulty with quantification over infinite domains. The first component has to do with a potentialist conception

---

[34]For a thorough analysis of Poincaré's contribution to the debate on predicativity see Heinzmann (1985). See also Feferman (2005) for a survey on predicativity.

of infinity that is at the heart of Poincaré's discussion. According to Poincaré, infinity is unboundedness, or the possibility of extending a finite set at will. Furthermore, the case of infinite domains is importantly different from the case of finitary domains. This contrasts with a common understanding of Cantorian set theory: the analogy between finite and infinite sets is often taken to be at the heart of Cantorian set theory's acceptance of actual infinity. Poincaré rejected actual infinity and thought that the illegitimate assumption of actual infinity gave rise to well-known paradoxes such as Russell's: we treat an infinite (and hence unbounded) set as if it had been completed. Let us consider once more the example above of the definition of X. As X is defined in terms of a relation with all of the G's, it seems that we would need to fix G's extension in order to make sense of the relation R between X and all the G's. However, for Poincaré, if G's extension is an infinite set, this becomes problematic. The thought seems to be that in the finitary case we may assume that the elements of G's extension are already available somehow independently of the defining condition G, so that it makes sense to think of the set as completed. Therefore fixing G's extension would seem to be in general unproblematic. However, when dealing with an infinite set, we cannot presuppose the availability of all the elements of the set independently of its definition: for any element of the set there is always a "next" element of it. This is because of the unbounded nature of infinite sets. The consequence is that a finitary defining condition is now the only tool we have for characterising the set; an enumeration of all its elements is not an option. In the case of an infinite domain, therefore, we have no more than the definition of the set to help us making sense of quantification over it: we cannot presuppose that all the elements of the set are already available and the set has been completed. To further clarify this point, it is useful to consider the second component of Poincaré's analysis of impredicativity, the concept of set.

The concept of set underlying Poincaré's reflection is very different from the familiar concept of set of ZF set theory. One may consider Poincaré's concept of set as essentially a variant of the traditional concept of set as extension of a concept.[35] Simplifying considerably, in the following, I take a set to be the *extension* of a definition or a condition (i.e. the collection of all the entities that satisfy that definition or condition). If we take this concept of set and combine it with a potentialist view of infinity as described above, we are brought to consider infinite sets whose extension

---

[35] See Parsons (2002).

is given exclusively in terms of the defining condition, say G. More specifically, an infinite set is not obtained by separating or selecting all the elements of the set from a given universe of sets; it arises from the possibility of always extending an initial finite fragment of it. Now, to make sense of quantification over an infinite set, its definition needs to be sufficiently informative. In the example of X and G above, G itself needs to provide enough information so that we can define X by postulating a relation R between X and *all* of the G's. This may be expressed by saying that G needs to offer a stable or *invariant* definition of its extension, one which determines once and for all what belongs to the set and which relations hold between the elements of the set. The possibility of 'new' elements appearing at a 'later stage' and 'disordering' the set should not occur. Poincaré (1912) hints at a genetic definition of sets, starting from a description of some initial elements, which then are used to construct (the description of) new ones, and so on. In this way, we definitely prescribe what belongs to a set, and the possibility of subsequently modifying the extension of its definition does not arise. In contemporary logical terms, going beyond Poincaré's remarks, we may say that a definition of an infinite set needs to give an *invariant* or absolute specification of all of its elements, starting from uncontroversial initial elements. If we have such an invariant specification of a set, generalisation over it becomes meaningful even if the set is infinite, since each element of the set can be specified (at least in principle) and no 'new' elements will accidentally slip in and disorder the set.

9. Back to Martin-Löf's type theory

Let us go back now to Martin-Löf's type theory. Here sets are defined by introduction and elimination rules which specify what a set is in terms of its elements (and give suitable identity conditions). It is tempting to think that sets in type theory make Poincaré's suggestion of genetic definition of a set from its elements fully precise: a set's definition gives not only a general membership condition, say G, that all and only the elements of G satisfy, but it also gives rules that specify its canonical elements. For example, the most fundamental infinite set, the natural numbers set, is given by introduction rules that describe the canonical elements of this set: 0 and the successor of any element of the natural numbers set.[36]

---

[36] The real numbers can also be represented in type theory. For example, by applying the rules for the constructor Σ (and other constructors) to the natural number set, one can express the notion of set of Cauchy real numbers.

Let us see now how quantification works in MLTT, including the case of infinite domains, such as the set of all the natural numbers. Given the Curry-Howard correspondence, quantification over a set is obtained by an application of the rules for the generalised Cartesian product, Π (universal quantifier), and the generalised disjoint union, Σ (existential quantifier). For example, a universally quantified statement of the form (∀x∈N) B(x) arises as particular reading of the generalised Cartesian product Π(N,B). An element of this type is a function that for each element n of N produces an element of B(n). By the Curry-Howard correspondence, this is a function that transforms a proof that n is an element of N (a natural number) to a proof of B(n). Martin-Löf (1984, 34) stresses the agreement of this understanding of the universal quantifier with the usual BHK interpretation of the quantifiers, according to which a proof of the above statement is a *method* which takes an *arbitrary* element n of N into a proof of B(n). Now, this reading of the universal quantifier does not require the prior full determination of the domain of quantification, N. For quantification to make sense, we do not need to presuppose that the extension of the condition defining a set (in this case the set of natural numbers) be fully determined. All what is required is a uniform method that taken an *arbitrary* element n of N produces a proof of B(n). This would suggest that we can make sense, intuitionistically, of a universal statement, even in the case in which the domain is potentially infinite.

Conclusion

In this article, I have reviewed a traditional picture of set theory, often implicit in introductory expositions of ZF set theory, for which classical first-order logic is taken for granted, and a collection of specific set-theoretic axioms is introduced 'on top' of it. I have argued that a comparison with constructive theories of sets questions the very possibility of drawing a clear demarcation between logical and set-theoretic principles. I have discussed two distinct cases: first of all the case of set theories, like CZF, which share their language with ZF, and secondly the case of MLTT. In the first case, the surprising observation is that the axioms of foundation and choice, which appear at first sight to be squarely set-theoretic, also act as logical principles.

In the case of MLTT, one witnesses an even stronger form of interaction between logical and set theoretic notions, in the form of the Curry-Howard isormorphism which identifies propositions with sets. This identification is at the heart of MLTT's predicativity and suggests an entanglement of logic and set theory. I have argued that a worry about classical quantification over infinite domains appears to be one of the motivations for MLTT's use of intuitionistic logic, and, consequently, its compliance with the Curry-Howard isomorphism. This bears similarities to the traditional intuitionist rejection of classical quantification over infinite sets. For the intuitionist, mathematics is primarily concerned with the infinite, and the latter requires the use of intuitionistic logic. I have noted that this profoundly differs from a traditional conceptions of classical logic as universally applicable and topic neutral.

A concern for generalisations over infinite domains is also at the heart of Poincaré's discussion on predicativity. Poincaré presents us with a potentialist view of infinity and combines it with a concept of set as dependent on a definition. This brings him to require a regimentation of definitions in mathematics, to avoid problematic impredicative definitions. The French mathematician hints at a genetic definition of sets by a step by step construction from its elements. I have suggested that MLTT may be seen as making precise sense of Poincaré's suggestion of predicative and genetic construction of sets from their elements.

To conclude, the above analysis questions not only the traditional view of classical logic as universally applicable and topic neutral, but the very possibility of clearly demarcating logic from set theory. The entanglement between logic and set theory prompted by constructive set theories deserves to be investigated in greater depth. Its analysis is bound to shed light on the relation between logic and sets beyond the constructive case.

Acknowledgements

The author is grateful to the editors for their kind invitation to contribute to this issue and for their patience. The author thanks the referees for their helpful comments and Andrea Cantini for commenting on a draft of this article.

Disclosure statement

No potential conflict of interest was reported by the author.

References

Aczel, P. 1978. "The Type Theoretic Interpretation of Constructive set Theory." In *'Logic Colloquium '77'*, edited by A. MacIntyre, L. Pacholski, and J. Paris, 55–66. Amsterdam-New York: North-Holland.

Aczel, P., and M. Rathjen. 2008. *Notes on constructive set theory*. June 2008.

Barendregt, H. P. 1981. *Lambda Calculus: Syntax and Semantics*, Vol. 103 of Studies in Logic and the Foundations of Mathematics, North–Holland.

Beeson, M. 1985. *Foundations of Constructive Mathematics*. Berlin: Springer Verlag.

Bishop, E. 1967. *Foundations of Constructive Analysis*. New York: McGraw-Hill.

Bishop, E., and D. S. Bridges. 1985. *Constructive Analysis*. Berlin: Springer.

Bridges, D. S., and E. Palmgren. 2013. "Constructive Mathematics." In The Stanford Encyclopedia of Philosophy, edited by E. N. Zalta, winter 2013 edn.

Bridges, D. S., and F. Richman. 1987. *Varieties of Constructive Mathematics*. Cambridge: Cambridge University Press.

Brouwer, L. E. J. 1912. "Intuitionism and Formalism." inaugural address at the University of Amsterdam, read October 14, 1912. Translated in Benacerraf and Putnam (1983).

Church, A. 1940. "A Formulation of the Simple Theory of Types." *Journal of Symbolic Logic* 5: 56–68.

Coquand, T. 1985. *Une théorie des constructions*, Thèse de troisième cycle, Paris VII.

Coquand, T. 1986. "An Analysis of Girard's Paradox", LICS, 227–236.

Coquand, T. 1989. "Metamathematical Investigations of a Calculus of Constructions." Technical Report 088; INRIA.

Coquand, T., and G. Huet. 1986. "The Calculus of Constructions", Technical Report RR-0530, INRIA.

Crosilla, L. 2015. "Set Theory: Constructive and Intuitionistic ZF." *The Stanford Encyclopedia of Philosophy* (Summer 2015 Edition), Edward N. Zalta (ed.). URL = <https://plato.stanford.edu/archives/sum2015/entries/set-theory-constructive/>.

Crosilla, L. 2017. "Predicativity and Feferman." In *Feferman on Foundations*, 423–447.

Dummett, M. 1991. *Frege: Philosophy of Mathematics*. Cambridge, MA: Harvard University Press.

Dummett, M. 1993. "What is Mathematics About?" In *The Seas of Language*, edited by A. George, 429–445. Oxford: University Press. Reprinted in Jacquette (2001), 19–30.

Feferman, S. 2005. "Predicativity." In *Handbook of the Philosophy of Mathematics and Logic*, edited by S. Shapiro, 590–624. Oxford: Oxford University Press.

Frege, G. 1953. *The Foundations of Arithmetic*. Translated by J. L. Austin. Oxford: Blackwell.

Friedman, H. 1973. "The Consistency of Classical set Theory Relative to a set Theory with Intuitionistic Logic." *Journal of Symbolic Logic* 38: 315–319.

Girard, J. 1972. "Interprétation fonctionnelle et élimination des coupures de l'arithmétique d'ordre supérieur," Thèse d'État, Paris VII.

Heinzmann, G. 1985. *Entre intuition et analyse: Poincaré et le concept de prédicativité*. Paris: Blanchard.

Hrbacek, K., and T. Jech. 1999. *Introduction to Set Theory*. 3rd ed. New York - Basel: Marcel Dekker, Inc.

Jacobs, B. 1989. "The Inconsistency of Higher Order Extensions of Martin-Löf's Type Theory." *Journal of Philosophical Logic* 18 (4): 399–422.

Martin-Löf, P. 1975. "An Intuitionistic Theory of Types: Predicative Part." In 'Logic Colloquium 1973', edited by H. E. Rose and J. C. Shepherdson, 73–118. Amsterdam: North–Holland.

Martin-Löf, P. 1982. "Constructive Mathematics and Computer Programming." In *Logic, Methodology, and Philosophy of Science VI*, edited by L. J. Choen, 153–175. Amsterdam: North–Holland.

Martin-Löf, P. 1984. *Intuitionistic Type Theory*. Naples: Bibliopolis.
Martin-Löf, P. 1987. "Truth of a Proposition, Evidence of a Judgment, Valididy of a Proof." *Synthese* 73: 407–420.
Martin-Löf, P. 2006. "100 Years of Zermelo's Axiom of Choice: What was the Problem with it?" *The Computer Journal,* 49 (3): 345–350.
Martin-Löf, P. 2008. "The Hilbert–Brouwer Controversy Resolved?" In *One Hundred Years of Intuitionism (1907–2007)*, edited by E. A. van Atten, 243–256. Basel: Publications des Archives Henri Poincaré.
Myhill, J. 1975. "Constructive set Theory." *J. Symbolic Logic* 40 (3): 347–382.
Nordström, B., K. Petersson, and J. M. Smith. 1990. *Programming in Martin-Löf's Type Theory: An Introduction*. Oxford: Oxford University Press.
Parsons, C. 2002. "Realism and the Debate on Impredicativity, 1917–1944." In *Reflections on the Foundations of Mathematics: Essays in Honor of Solomon Feferman*, edited by W. Sieg, R. Sommer, and C. Talcott, 372–389. Association for Symbolic Logic.
Poincaré, H. 1905. "Les mathématiques et la logique." *Revue de Métaphysique et de Morale* 1: 815–835.
Poincaré, H. 1906. "Les mathématiques et la logique." *Revue de Métaphysique et de Morale* 14: 294–317.
Poincaré, H. 1909. "La logique de l'infini." *Revue de Métaphysique et de Morale* 17: 461–482.
Poincaré, H. 1912. "La logique de l'infini." *Scientia; Rivista Di Scienza* 12: 1–11.
Russell, B. 1906a. "Les paradoxes de la logique." *Revue de Métaphysique et de Morale* 14: 627–650.
Russell, B. 1906b. "On Some Difficulties in the Theory of Transfinite Numbers and Order Types." *Proceedings of the London Mathematical Society* 4: 29–53.
Russell, B. 1908. "Mathematical Logic as Based on the Theory of Types." *American Journal of Mathematics* 30: 222–262.
Sambin, G., and S. Valentini. 1998. "Building up a Toolbox for Martin-Löf Type Theory: Subset Theory." In *Twenty-Five Years of Constructive Type Theory*, edited by G. Sambin and J. Smith, 206–221. Oxford: Oxford University Press. Oxford Logic Guides, 36.
Sundholm, G. 1986. "Proof Theory and Meaning." In *Handbook of Philosophical Logic*. Vol. III., edited by D. Gabbay and F. Guenthner, 471–506. Dordrecht: Springer.
Sundholm, G. 1994. "Existence, Proof and Truth-Making: A Perspective on the Intuitionistic Conception of Truth." *Topoi* 13: 117–126.
Suppes, P. 1960. *Axiomatic Set Theory*. The University Series in Undergraduate Mathematics, D. Van Nostrand Co., Inc., Princeton, N.J.-Toronto-London-New York xii+265 pp.
Troelstra, A. S., and D. van Dalen. 1988. *Constructivism in Mathematics: An Introduction*. 2 Vols. Amsterdam: North Holland.
Weyl, H. 1918. *Das Kontinuum. Kritische Untersuchungen über die Grundlagen der Analysis*. Leipzig: Veit.
Weyl, Hermann. 1946. "Mathematics and Logic." *The American Mathematical Monthly* 53 (1): 2–13.

Critical plural logic [1]

Salvatore Florio and Øystein Linnebo

ABSTRACT
What is the relation between some things and the set of these things? Mathematical practice does not provide a univocal answer. On the one hand, it relies on ordinary plural talk, which is implicitly committed to a traditional form of plural logic. On the other hand, mathematical practice favors a liberal view of definitions which entails that traditional plural logic must be restricted. We explore this predicament and develop a "critical" alternative to traditional plural logic.

1. Introduction

English and other natural languages contain plural expressions, which allow us to talk about many objects simultaneously, for example:

(1) The students cooperate.
(2) The natural numbers are infinite.

How should such sentences be analyzed? A common strategy does without plurals and relies on sets.[2] Plural talk is eliminated in favor of singular talk about sets. For example, sentence (2) is analyzed as:

(3) The set $\{x : x$ is a natural number$\}$ is infinite.

In recent years, there has been a surge of interest in an alternative strategy that makes uses of plural logic. This is a logical system that takes plurals at face value. When analyzing language, there is no need to eliminate the plural resources of English in favor of talk about sets or any other singular

---

[1] This article draws on material from our forthcoming book The *Many and the One: A Philosophical Study*, which is a comprehensive study of the logic, meaning, and metaphysics of plurals. Here we follow a particular thread concerned with the relation between pluralities and sets, relying especially on Chapters 2, 4, and 12. For useful comments and discussion, we would like to thank two anonymous referees, Jose Ferreiros, Peter Fritz, Simon Hewitt, David Nicolas, Alex Oliver, Agustin Rayo, Sam Roberts, Stewart Shapiro, Timothy Smiley, Eric Snyder, Hans Robin Solberg, and Gabriel Uzquiano, as well as audiences at the Fifth International Meeting of the Association for the Philosophy of Mathematical Practice in Zurich and the International Conference for Philosophy of Science and Formal Methods in Philosophy in Gdanisk.
[2] See, e.g., [Quine, 1982] and [Resnik, 1988].

resources. Rather, the plural resources can be retained as primitive, not understood in terms of anything else.[3]

Assuming the alternative strategy is right, we have (at least) two different ways to talk about many objects simultaneously: plurally and by means of singular talk about sets. What is the relation between these two ways? That is, what is the relation between some things and the set of these things? In this article, we are especially interested in how mathematical practice bears on these questions. At the center of our discussion are:

(i) Cantor's and Gödel's appeals to plurals to explain the notion of a set;
(ii) a liberal view of mathematical definitions, espoused by Cantor and others, which entails that every plurality defines a set.

As we explain, this liberal view requires us to replace the traditional logic of plurals with a more "critical" alternative.

Two larger questions pervade our discussion. The first question concerns how, and on what basis, we should choose a "correct" logic — in this particular case, a logic of plurals. We argue that the choice of a plural logic is entangled with some hard questions in the philosophy and foundations of mathematics. The second larger question concerns what, and how, philosophers can learn from studying mathematical practice, which is not always internally consistent. On the one hand, it is implicitly committed to a traditional form of traditional plural logic, at least in so far as this practice relies on ordinary plural language. On the other hand, mathematical practice favors a liberal view of definitions which entails that traditional plural logic must be restricted. We must therefore be extremely careful when attempting to extract philosophical lessons from mathematical practice. A detailed analysis of a broadly philosophical character is needed to adjudicate between the conflicting features of the practice.

2. Plural Logic

Let us begin by describing a language that may be used to regiment a wide range of natural language uses of plurals and to represent many valid patterns of reasoning that essentially involve plural expressions. This language is associated with what is known in the philosophical literature as PFO+, which is short for *plural first-order logic plus plural predicates*. In one variant or another, it is the most common regimenting language for plurals in philosophical logic.[4]

---

[3] This surge draws much of its inspiration from seminal work by George Boolos [1984; 1985], but Russell's [1903] notion of a "class as many" is an important anticipation.

[4] We adopt the notation for variables used in [Rayo, 2002] and [Linnebo, 2003]. An ancestor of this notation is found in [Burgess and Rosen, 1997]. Other authors represent plural variables by means of different typographical conventions: boldface letters [Oliver and Smiley, 2016, pp.

We start with the the standard language of first-order logic and expand it by making the following additions.

A. Plural terms, comprising plural variables (*vv, xx, yy,...*, and variously indexed variants thereof) and plural constants (*aa, bb,...*, and variants thereof), roughly corresponding to the natural language pronoun 'they' and to plural proper names, respectively.
B. Quantifiers that bind plural variables ($\forall vv$, $\exists xx$,...).
C. A binary predicate $\prec$ for plural membership, corresponding to the natural language 'is one of' or 'is among'. This predicate is treated as logical.
D. Symbols for collective plural predicates with numerical superscripts representing the predicate's arity ($P^1, P^2,..., Q^1,...$, and variously indexed variants thereof). Examples of collective plural predicates are '...cooperate', '... gather', '... meet', '... outnumber...', '...are infinite'. For economy, we leave the arity unmarked.

We use this language to define two important many-many relations: plural inclusion and plural identity. Plural inclusion, expressed by 'are among', is symbolized as '$\preceq$' and defined thus:

$$xx \preceq yy \leftrightarrow_{def} \forall z(z \prec xx \rightarrow z \prec yy).$$

That is, these things are among those things just in case every one of these is one of those. Plural identity (symbolized as '≈') can be defined as mutual plural inclusion. In symbols:

$$xx \approx yy \leftrightarrow_{def} (xx \preceq yy \wedge yy \preceq xx).$$

That is, two pluralities are identical just in case they are coextensive.
To illustrate the use of PFO+, let us provide some examples of regimentation based on this language.

(4) Some students cooperated.
(5) $\exists xx (\forall y(y \prec xx \rightarrow S(y)) \wedge C(xx))$.
(6) Bunsen and Kirchhoff laid the foundations of spectral analysis.
(7) $\exists xx (\forall y(y \prec xx \leftrightarrow (y = b \vee y = k)) \wedge L(xx))$.
(8) Some critics admire only one another.
(9) $\exists xx (\forall x(x \prec xx \rightarrow C(x)) \wedge \forall x \forall y[(x \prec xx \wedge A(x, y)) \rightarrow (y \prec xx \wedge x \neq y)])$.

The formal system PFO+ comes equipped with logical axioms and rules of inference aimed at capturing correct reasoning in the fragment of natural

---

221–223], capitalized letters [McKay, 2006, pp. 59–60], or singular variables pluralized with an 's' [Yi, 1999, pp. 177–178]

language that is being regimented. The axioms and rules associated with the logical vocabulary of ordinary first-order logic are the usual ones. For example, one could rely on introduction and elimination rules for each logical expression. The plural quantifiers are governed by axioms or rules analogous to those governing the first-order quantifiers.

Plural logic is often taken to include some further, very intuitive axioms. First, every plurality is non-empty:

(Non-empty) $\forall xx \, \exists yy \prec xx$.

Then, there is an axiom scheme of indiscernibility stating that coextensive pluralities satisfy the same formulas:

(Indiscernibility) $\forall xx \, \forall cyy[xx \approx yy \rightarrow (\varphi(xx) \leftrightarrow \varphi(yy))]$

(The formula φ may contain parameters. So, strictly speaking, we have the universal closure of each instance of the displayed axiom scheme. Henceforth, we assume this reading for similar axiom schemes.) This is a plural analogue of Leibniz's law of the indiscernibility of identicals, and as such, the scheme needs to be restricted to formulas $\varphi\,(xx)$ that do not set up intensional contexts.

Finally, there is the unrestricted axiom scheme of plural comprehension, an intuitive principle that provides information about what pluralities there are. Informally, for any formula $\varphi\,(x)$ containing x but not xx free, we have an axiom stating that if $\varphi\,(x)$ is satisfied by at least one thing, then there are the things each of which satisfies $\varphi\,(x)$:

(P-Comp) $\exists x\varphi(x) \rightarrow \exists xx \, \forall x(x \prec xx \leftrightarrow \varphi(x))$.

We refer to an axiomatization of plural logic based on the principles just described as traditional plural logic. This is to emphasize its prominence in the literature. We believe traditional plural logic is implicit in our ordinary use of plural language — and thus also in mathematical practice, which sometimes relies on such language.

Philosophers often proceed to make strong claims about this deductive system: it is aptly and rightly called "plural logic", because its principles have the same privileged status as is widely accorded to ordinary first-order logic.[5] Thus, plural logic is pure logic, not set theory in disguise. Although this is not the place for a thorough discussion of what counts as pure logic, we wish to make some brief remarks, focusing on three features.

One aspect of logicality is *topic neutrality*. This is based on the simple, intuitive idea that logical principles should be applicable to reasoning about any

---

[5] See, e.g., [Boolos, 1985] and [Hossack, 2000].

subject matter. By contrast, non-logical principles are only applicable to particular domains. The laws of physics, for instance, concern the physical world and do not apply when reasoning about natural numbers or other abstract entities. Plural logic seems to satisfy this intuitive notion of topic neutrality: the validity of the principles of plural logic does not appear to be restricted to specific domains. As partial evidence for the topic neutrality of plural logic, one may point out that, when available, pluralization as a morphological transformation does not depend in any systematic way on the kind of objects one speaks about; for example, both concrete and abstract count nouns exhibit plural forms. While we agree that plural logic *in some form* is topic-neutral, we deny that *traditional* plural logic has this status. Specifically, we argue that the unrestricted plural comprehension scheme is valid only for special kinds of domains — loosely speaking, domains that are properly circumscribed.

A second aspect of logicality is *ontological innocence*: a logical truth should not carry any ontological commitments. (There is usually one exception: the existence of at least one object. But even this commitment is generally tolerated only so as to streamline logical theory, not endorsed for doctrinal reasons.) Plural logic is widely held to be ontologically innocent in this sense.[6] The plural existential quantifier expresses that there is one or more objects *of the sort to which one is already committed*; it does not introduce any new ontological commitment. In other work, however, we dispute this widely held view, arguing that there is an important sense in which plural logic does carry non-trivial commitments of a broadly ontological sort.[7] We choose not to enter into this debate here.

A third aspect of logicality is the idea that logical notions and principles permit a special kind of epistemic primacy. Logical notions can be grasped without relying on non-logical notions. Likewise, logical truths, if knowable, can be known independently of non-logical truths, including those of mathematics. The claim that the principles of plural logic enjoy this kind of epistemic primacy will be challenged in what follows. As advertised, we take plural logic to be entangled with certain broadly set-theoretic principles.

3. Plural logic vs a simple set theory: a formal comparison

What is the relation between some things and their set? Let us begin with a formal comparison of plural logic and set theory, which will clarify some important technical aspects of the question. In later sections, we will address some philosophical issues concerning the relation between some things and their set.

---

[6] Defenses of the innocence of plural logic are put forth by, among others, Boolos [1984; 1985], Yi [1999; 2002; 2005; 2006], Hossack [2000], Oliver and Smiley [2001; 2016], Rayo [2002], and McKay [2006].
[7] See [Florio and Linnebo, 2016] and [Florio and Linnebo, forthcoming, Chap. 8].

Assume we start with an ordinary singular first-order language whose quantifiers range over certain objects. Let us refer to these objects as *individuals*. We are interested in ways to talk simultaneously about many individuals. The most familiar option, at least to anyone with some training in mathematics, is to use set theory. A set is a single object that has zero or more *elements*. Talking about a single set thus provides a way to talk about all of its elements simultaneously. We can, for example, convey information about Russell and Whitehead by talking about their set {Russell, Whitehead}. The information that they are philosophers can be conveyed by saying that every element of the set is a philosopher. Suppose, more generally, that we want to talk about the φ's, where φ is a formula of our language that is satisfied by at least one object. According to the present strategy, this can be achieved by talking about the associated set, namely $\{x : \varphi(x)\}$.

However, it is not obvious that this strategy always works. After all, the lesson of the set-theoretic paradoxes is that not every formula defines a set. (We assume classical logic.) The most famous example is Russell's paradox of the set of all sets that are not elements of themselves. Consider the formula that serves as a condition for membership in this would-be set: $x \notin x$. Suppose this formula defines a set R. Now ask: is R an element of itself? The answer is affirmative if and only if R satisfies the membership condition. In other words: $R \in R$ if and only if $R \notin R$. But this is a contradiction!

Thankfully, the problem posed by the set-theoretic paradoxes can be put off, at least for a little while. The paradoxes do not arise when we consider only sets of individuals drawn from a fixed first-order domain. And for present purposes, this is all we need. Let us therefore consider a very simple set theory, which satisfies our present needs but does not give rise to paradoxes.

We need to distinguish between individuals and sets of individuals. To do so, it is convenient to use a *two-sorted* language. Such languages are easily explained because they are implicit in various mathematical practices. For example, in geometry we often use one set of variables to range over points (say, $p_1, p_2, ...$) and another set of variables to range over lines (say, $l_1, l_2, ...$). We adopt a similar approach to our simple set theory, letting lower-case variables range over individuals ($x, y, ...$) and upper-case variables ($X, Y, ...$) range over sets of individuals. We refer to these as *individual variables and set variables*, respectively. If desired, we can add constants of either sort. There are sortal restrictions on the formation rules. For instance, the language has a membership predicate '$\in$' whose first argument can only be an individual term and whose second argument can only be a set term. Thus, '$a \in X$' means that the individual a is an element of the set X. In addition to the ordinary identity predicate, which can be flanked by any two individual terms, our extended language contains a set-identity predicate, which can be flanked only by set terms. For convenience, we write this predicate as the usual identity sign. Given the restrictions just mentioned, it is impermissible to make identity claims involving both an individual and a set term (such as '$a = X$'). We call this language $\mathcal{L}_{SST}$, and we let $\mathcal{L}_{SST}^+$ be the language obtained from $\mathcal{L}_{SST}$ by adding predicates that take set terms as arguments. This is an optional extra to which we will return.

We now formulate our simple set theory, SST, based on the axioms and rules of two-sorted classical logic. First, we adopt the axiom of extensionality for sets:

(S-Ext) $\quad\quad \forall x(x \in X \leftrightarrow x \in Y) \rightarrow X = Y$.

Then, we adopt an axiom scheme of set comprehension:

(S-Comp) $\quad\quad \exists X \, \forall \, x(x \in X \leftrightarrow \varphi(x))$,

where X does not occur free in φ. The theory SST+ is obtained by adapting the axioms and rules of SST to the richer language $\mathcal{L}_{SST}^+$. Notice how Russell's paradox is blocked by the use of separate sorts for individuals and their sets. In our two-sorted language, the membership condition for the offending set, namely $x \notin x$, cannot even be formulated. The same can be seen to hold for the other set-theoretic paradoxes, such as Burali-Forti.

Consider a domain of individuals to which both plural logic and the simple set theory are applicable. We thus have two different ways to talk about many objects simultaneously. As we will now show, these two different ways of talking share a common structure.

To begin with, the two languages share a common stock of variables x_i that take as their values one individual at a time. Further, each language has an additional stock of variables that are used to convey information about (loosely speaking) collections of individuals: plural variables xx_i, which take as their values many individuals simultaneously, or set variables X_i, which take as their values a single set of individuals. Finally, each language has a predicate for membership in a collection: $x_i \prec xx_j$ for "x_i is one of xx_j" or $x_i \in X_j$ for "x_i is an element of X_j". These observations suggest that it should be straightforward to translate back and forth between the two languages. One can simply replace $\prec$ with x_i and xx_j with X_j, and *vice versa*.

However, there are two wrinkles to be ironed out:

- $\mathcal{L}_{SST}$ has an identity predicate that can be flanked by set terms, whereas the language of PFO has no identity predicate that can be flanked by plural terms.
- SST postulates an empty set, whereas PFO has an axiom stating that every plurality is non-empty.

Fortunately, both problems are easily overcome. It is possible to define a translation from each language to the other such that each sentence and its translation convey the same information, at least as far as the individuals are concerned. The only difference is that one sentence does so by utilizing plural resources, while the other uses set-theoretic resources.[8]

The translations satisfy the following important conditions:

(i) Each translation is recursive; that is, there is an effective algorithm for carrying out the translation.
(ii) Each translation respects logical structure; for example, the translation of a negation $\neg \varphi$ is the negation of the translation of φ.
(iii) Every axiom of each of the two theories is translated as a theorem of the other theory; for example, each axiom of PFO is translated as a theorem of SST.

More generally, let τ be a translation from the language of one theory T_1 to that of another theory T_2 such that these three conditions are satisfied. Then τ is said to be an *interpretation* of T_1 in T_2. Thus, we have that each of our two theories PFO and SST can be interpreted in the other, and likewise with PFO+ and SST+.

It is important to be absolutely clear about what the mutual interpretability of two theories does and does not establish. Interpretability is a purely formal notion: it concerns a translation preserving theoremhood, and it allows us to turn a model of one theory recursively into a model of another. So, two mutually interpretable theories are equivalent for the purposes of formal logic. However, there is no guarantee that the equivalence will extend beyond those purposes. Suppose the two languages are meaningful. Then there is no guarantee that the translation preserves the kinds of extra-logical properties that philosophers often care about. For example, the translation need not preserve features of sentences such as:

- truth value;
- meaning (perhaps understood as the set of possible worlds at which a sentence is true);
- epistemic status (such as apriority or aposteriority);
- ontological commitments.

It is often controversial whether a translation preserves these features. The translations we consider here are no exception.

4. Should sets or pluralities be eliminated?

What is the significance of the shared structure, or mutual interpretability, that we just observed? Is this *merely* a technical result? Or does the

---

[8] This observation is due to Boolos [1984]. For an exposition, see [Florio and Linnebo, forthcoming, Appendix 4.A].

technical result have some broader philosophical significance? When the structure of one theory can be recovered within that of another, this raises the question of whether one of the theories can be eliminated in favor of the other. In the present context, there are three options. First, one may eliminate pluralities in favor of sets. Second, one may proceed in the opposite direction and eliminate sets in favor of pluralities. Finally, one may refrain from any elimination and retain both pluralities and sets. All three options have their defenders. Let us consider them in turn.

First, some philosophers seek to eliminate sets in favor of pluralities. That is, we can and should interpret ordinary 'set' talk without relying on set-theoretic resources. A classic paper by Black [1971] can be read as advocating this view.[9] More recently, Oliver and Smiley [2016, pp. 316–317] have expressed considerable sympathy for the view, claiming at least to have shifted the burden of proof onto its opponents. Black observes that ordinary language often talks about sets: expressions such as 'my set of chessmen' or 'that set of books' feel fairly natural to English speakers. By reflecting on ordinary uses of the word 'set', he argues, we can come to see the intimate connection between talk about a set and about its elements. More specifically, we can come to realize that basic uses of the word 'set' are simply substitutes for plural expressions such as lists of terms or plural descriptions. In his example, the sentence 'a certain set of men is running for office' is what he calls an "indefinite surrogate" for the statement that, say, Tom, Dick, and Harry are running for office [Black, 1971, p. 631].

Second, other philosophers hold that the plural locutions found in English and other natural languages should be eliminated in favor of talk about sets. Quine and Resnik are advocates of this view.[10] For Quine at least, this is at root a claim about regimentation into our scientific language. It is indisputable that many natural languages contain plural locutions. But our best scientific theory of the world has no need for such locutions. This theory is to be formulated in a singular language whose quantifiers also range over sets. When regimenting natural language into this scientific language, the plural locutions of the former should be analyzed by means of the set talk of the latter. In short, for scientific purposes, we should eschew plural resources and instead rely on set-theoretic resources. These resources also suffice to interpret "the vulgar" (as Quine once put it), that is, to regiment the plural resources indisputably found in English and other natural languages.

Finally, one may hold that neither system should be eliminated in favor of the other, because both plural logic and set theory are legitimate

---

[9] We should note that this is not the only way to read Black. It is not entirely clear whether he proposes an eliminative reduction or favors some form of non-eliminative reductionism. An eliminative proposal is developed by Hossack [2000], who appeals to plurals and plural properties to eliminate sets.

[10] See footnote 1.

and earn their keep in our best scientific theory. Following Cantor and and Gödel, this is the view that we will defend. Suppose we are right that both systems should be retained. This gives rise to some further questions concerning their relation. We will be particularly concerned with two such questions.

(i) Every non-empty set obviously corresponds to a plurality, namely the elements of the set. What about the other direction? Does every plurality correspond to a set? If not, under what conditions do some things form a set?[11]

(ii) Suppose that some objects form a set. Can these objects be used to shed light on, or give an account of, the set that they form?

First, however, let us explain why we reject both of the eliminative proposals.

Why not eliminate sets in favor of pluralities, following Black and others? Black recognizes that there is a gap between ordinary uses of the word 'set' and its uses in mathematics. For instance, ordinary speakers untrained in abstract mathematics often have misgivings about the empty set. If sets are collections of things, how can there be a collection of nothing whatsoever? Despite such misgivings, Black contends that we can rely on our ordinary understanding of plurals to make sense of "idealized" uses of the word 'set' as it occurs in mathematics.

However, there is an obvious difficulty for Black's contention. Talk of *sets of sets* is ubiquitous in mathematics and, as we will see shortly, such "nested" sets are essential to the now-dominant iterative conception of set. How can we account for these uses of the word 'set'? If talk about sets is shorthand for talk about pluralities, then sets of sets would seem to correspond to higher-order pluralities, that is, "pluralities of pluralities".[12] It is controversial whether such higher-order pluralities make sense, but a putative example is given in following sentence.

(10) My children, your children, and her children competed against each other.

The subject of this sentence appears to be a "nested" plural, that is, a plural expression formed by combining three other plural expressions. Arguably, this nesting of the subject is semantically significant. The claim is not merely that all the children in question compete against each other but that they do so in teams, each team comprising the children of each parent.[13]

---

[11] See [Hewitt, 2015] for a useful overview of this issue.
[12] For proposals along these lines, see [Simons, 2016] and [Oliver and Smiley, 2016, Chap. 15].
[13] We discuss whether there are higher-order pluralities in [Florio and Linnebo, forthcoming, Chap. 10].

While the availability of higher-order pluralities is a necessary condition for the envisaged elimination of sets, we deny that it is sufficient. As observed, the language of mathematics talks extensively about sets and appears to treat these as objects. Other things being equal, it would be good to follow mathematical practice and take this language at face value. In the absence of a strong reason to deviate from the practice, this yields an independent reason not to eliminate sets. This reason is even stronger for those who accept other mathematical objects such as numbers. If numbers are accepted, why not also accept sets?

We turn now to Quine's and Resnik's suggestion that pluralities be eliminated in favor of sets. It is often objected that this form of elimination would give rise to paradoxes.[14] We do not find these arguments entirely convincing.[15] Instead of entering into this debate here, however, we wish to lay out another — and, we believe, more compelling — reason why pluralities should not be eliminated in favor of sets. The reason is simply that pluralities are needed to give an account of sets. If pluralities were eliminated in favor of sets, we could not use plural reasoning to give such an account. In sum, to retain an attractive account of sets in terms of pluralities, we cannot eliminate plurals.

5. Accounting for sets in terms of pluralities

What is the promised account of sets in terms of pluralities? It is useful to recall how Cantor, the father of modern set theory, sought to explain the concept of set.[16]

> By a 'manifold' or 'set' I understand in general every many which can be thought of as one, *i.e.*, every totality of determinate elements which can be bound together into a whole through a law [...].[17]

That is, a set is a "many thought of as one". Of course, it is far from clear how this is to be understood. But there can be no doubt that Cantor sought to understand a set in terms of the many objects that are its elements and that are somehow "thought of as one".[18]

---

[14] See, *e.g.*, [Boolos, 1984, pp. 440–444; Lewis, 1991, p. 68; Schein, 1993, Chap. 2, §3.3; Higginbotham, 1998, pp. 14–17; Oliver and Smiley, 2001, pp. 303–305; Rayo, 2002, pp. 439–440; McKay, 2006, pp. 31–32].
[15] See [Florio and Linnebo, forthcoming, §3.4].
[16] Unter einer Mannichfaltigkeit oder Menge verstehe ich nämlich allgemein jedes Viele, welches sich als Eines denken lässt, d.h. jeden Inbegriff bestimmter Elemente, welcher durch ein Gesetz zu einem Ganzen verbunden werden kann [...]. [Cantor, 1883,p.42]
[17] Since the exact choice of words is important to the point we are making here, we have chosen to provide our own translation of this passage and the next.
[18] Unter einer ‚Menge' verstehen wir jede Zusammenfassung M von bestimmten wohlunterschiedenen Objekten m unsrer Anschauung oder unseres Denkens (welche die ‚Elemente' von M genannt werden) zu einem Ganzen. In Zeichen drucken wir dies so aus: $M = \{m\}$.[Cantor, 1895, p. 481]

By a 'set' we understand every collection into a whole M of determinate, well-distinguished objects m of our intuition or our thought (which will be called the 'elements' of M). We write this as: M = {m}.

It is tempting to read Cantor's variable 'm' as a plural variable (see also [Oliver and Smiley, 2016, pp. 4–5]). So, in line with our notation, let us replace this variable with 'mm'. A set M is then said to be a collection into one of some well-distinguished objects mm, namely the elements of M. And we write M = {mm}. A closely related idea is endorsed by Gödel, who, in a passage to be considered shortly, discusses a "set of" operation that takes some "well-defined objects" to their set.

More generally, many philosophers and mathematicians believe that the elements of a set are somehow "prior to" the set itself and that the set is somehow "constituted" by its elements.[18][19] Assume xx form a set $\{xx\}$. Then the objects xx can be used to give an account of $\{xx\}$. That is, properties and relations involving the set are explained in terms of properties and relations involving the plurality of its elements. Why is a an element of $\{xx\}$? An answer immediately suggests itself: because a is one of xx! Why is $\{xx\}$ identical with $\{yy\}$? Again, the answer seems obvious: because xx are the very same objects as yy.

All of these remarks suggest to us a liberal view of mathematical definitions, which we will first sketch and then spell out and defend. This liberal view takes it to be sufficient for a mathematical object to exist that an adequate definition of it has been provided. The adequacy in question is understood as follows. Consider a "properly circumscribed" domain of objects standing in certain relations. We would like to define one or more additional objects. Suppose our definition determines the truth of any atomic statement concerned with the desired "new" objects by means of some statement concerned solely with the "old" objects with which we began. Then, according to the liberal view, the definition is permissible.

To illustrate the point, let us apply the view to the case of sets. Again, consider some properly circumscribed domain of objects. For every plurality of objects xx from this domain, we postulate their set $\{xx\}$, with the understanding that atomic statements concerned with any new sets should be assessed in the way mentioned above.

(i) $\{xx\} = \{yy\}$ if and only if $xx \approx yy$.
(ii) $a \in \{xx\}$ if and only if $a \prec xx$.

Notice how this account determines the truth of any atomic statement concerned with the "new" sets solely in terms of the "old" objects with which we began, as required by the liberal view.

---

[19] See, *e.g.*, [Parsons, 1977] and [Fine, 1991].

What about the empty set? Here there is a threat of a mismatch. While standard set theory accepts an empty set, traditional plural logic does not accept an empty plurality. But we are confident that this threat can be addressed. One option is to break with traditional plural logic and accept an empty plurality, perhaps on the grounds that, although this is not how plurals work in English and many other natural languages, there are coherent languages where plurals do behave in this way (see [Burgess and Rosen, 1997, pp. 154–155]). Another option is to break with standard set theory and abandon the empty set. However, we would prefer not to deviate from successful scientific practice, in this case set theory, unless there are compelling reasons to do so. Finally, an elegant option proposed (in a different context) by Oliver and Smiley [2016, pp. 88–89] is to allow "co-partial functions", that is, functions that can have a value even where the argument is undefined. Suppose the "set of" operation $xx \mapsto \{xx\}$ is such a function. Then, applied to an undefined argument, this function can have the empty set as its value.

Next, can we account for nested sets? This means going beyond the simple set theory discussed above to form a stronger set theory, where the threat of paradox re-emerges. The standard response to this threat is the so-called iterative conception of set. One of the first clear expressions of this conception is given in a famous passage by Gödel.[20]

> The concept of set, however, according to which a set is anything obtainable from the integers (or some other well-defined objects) by iterated application of the operation "set of", and not something obtained by dividing the totality of all existing things into two categories, has never led to any antinomy whatsoever; that is, the perfectly "naïve" and uncritical working with this concept of set has so far proved completely self-consistent. [Göidel, 1964, p. 180]

The passage calls for some explanation. First, Giodel distinguishes the iterative conception of set from a problematic conception based on the idea of "dividing the totality of all existing things into two categories". Consider a condition that any object may or may not satisfy. One might then attempt to use this condition to divide the totality of all objects into two sets: the set of objects that satisfy the condition and the set of those that do not. But this approach to sets is problematic: as we have seen, it gives rise to Russell's paradox.

By contrast, the iterative conception starts with the integers or "some other well-defined objects". We are then told to consider iterated applications of the operation "set of". An example will help. Assume we start, at what we may call stage 0, with two objects, say a and b. The "set of" operation can be applied to any plurality of objects available at stage 0 to form

---

[20] The passage contains some footnotes that we elide.

their set. Thus, at stage 1, which results from the application of this operation to the objects available at stage 0, we have the following sets: $\emptyset, \{a\}, \{b\}$, and $\{a, b\}$. So, at stage 1, we have six objects, namely a and b together with four sets that were not available at stage 0. Now we can apply the "set of" operation again, this time to the objects available at stage 1. This yields sets such as $\{\emptyset, a\}, \{\{a\}, \{b\}\}$, and many others. Note that, by this procedure, the objects available at any given stage form a set at the next stage.

There is a more systematic way to describe what takes us from one stage α to the next stage $\alpha + 1$. For any set S, let its powerset, $\wp(S)$, be the set of all subsets of S, that is:

$$\wp(S) = \{x : x \subseteq S\}.$$

Suppose the objects available at stage α are the elements of V_α. Then at stage $\alpha + 1$ we form all the subsets of V_α. So, at stage $\alpha + 1$, we have the elements of V_α as well as those of $\wp(V_\alpha)$. In symbols: $V_{\alpha+1} = V_\alpha \cup \wp(V_\alpha)$. Again, we have by this procedure that all the sets available at stage α, taken together, form a set at stage $\alpha + 1$.

In fact, we want to consider really long iterations of the "set of" operation. The first step is to define V_ω as the result of continuing in this way as many times as there are natural numbers. We do this by letting V_ω be the union of all of the collections V_n generated at a finite stage: $V_\omega = \bigcup_{n<\omega} V_n$. More generally, for any limit ordinal λ, we let V_λ be the union of all the collections of sets we have generated: $V_\lambda = \bigcup_{\gamma<\lambda} V_\gamma$. The *cumulative hierarchy of sets*, V, is the union of all of the V_α.

However, as Gödel observes in a footnote to the passage just quoted, V is not a set. There is no stage at which all sets are available to form a universal set. For any stage, there is a later stage containing even more sets. As a result, we ban the universal set and any other set that would lead to paradox. This raises the question of the status of the cumulative hierarchy itself, including the question of whether "it" even exists as an object.

6. Proper classes as pluralities

In fact, both the problem posed by the ontological status of the entire cumulative hierarchy and the proposed solution of invoking plurals generalize. Let us use the word 'collection' in an informal way for anything that has a membership structure, such as a set, class, plurality, or indeed even a Fregean concept — when the relation between instance and concept is regarded as a membership structure. We will now explain the apparent need to talk about collections that are too large to form sets, why these are sometimes regarded as problematic, and finally a brilliant proposal due to Boolos, namely that plural logic provides a way to make sense of these collections.

Let us begin with the need for a novel type of collection, in addition to sets. There are several reasons for this need. Boolos mentions two. First, collections are needed to make sense of the domain of set theory, namely the cumulative hierarchy V. For example, we would like to say that V is the subject matter of set theory and that V is well-founded.

Second, collections are needed to understand and justify two axiom schemes that are part of ordinary Zermelo-Fraenkel set theory, ZFC, namely Replacement and Separation.[21] Both of these take the form of an infinite family of axioms. Consider Separation. ZFC contains an axiom

(Sep) $\quad \forall z \exists y \forall x (x \in y \leftrightarrow x \in z \wedge \varphi)$

for each of the infinitely many formulas φ of its language. Behind this infinite lot of axioms lies a single, unified idea that can be expressed by reference to collections. For every collection C and every set x, there is a set y of all those elements of x that belong to C. Suppose we can quantify over collections. Then the infinitely many Separation axioms could be unified as the single axiom:[22]

(C-Sep) $\quad \forall C \forall x \exists y \forall z (z \in y \leftrightarrow z \in x \wedge z$ belongs to $C)$.

In the literature, the desired collections are often known as *classes*, some of which can be shown to be "too big" to be sets. These are called proper classes. But what would these *proper classes* be? Just like sets, they are collections of many objects into one. But why, then, are proper classes not sets? As Boolos [1984, p. 442] nicely observes, "[s]et theory is supposed to be a theory about *all* set-like objects".

Adding proper classes to a theory of sets is just like adding yet another layer of sets on top of the sets already recognized. In light of this, why should the proper classes not count as just more sets? William Reinhardt [1974, p. 122] puts the point well:

> [O]ur idea of set comes from the cumulative hierarchy, so if you are going to add a layer at the top it looks like you forgot to finish the hierarchy.[23]

Plural logic seems to provide precisely what we need. There is no need for a proper class to be a *single object* that somehow collects together many things into one. Instead of referring in a singular way to a proper class,

---

[21] Analogous considerations apply to the arithmetical principle of induction.
[22] See also [Kreisel, 1967].
[23] For a useful elaboration of the point, see [Maddy, 1983, p. 122].

construed as an object, why not simply refer plurally to its many members? In this way, we eliminate talk about proper classes in favor of plural talk about their members. For example, there is no need for the cumulative hierarchy to be an object. It suffices to talk plurally about *all the sets*. Next, consider the axiom scheme of Separation. This can now be given a single uniform plural formulation. Given any objects *pp* and any set *x*, there is a set *y* of precisely those elements of *x* that are also among pp:

(P-Sep) $\forall pp \, \forall x \, \exists y \, \forall z (z \in y \leftrightarrow z \in x \wedge z \prec pp)$.

Of course, to represent all of the classes that we might be interested in, we would need an unrestricted form of plural comprehension.

Let us take stock. We have described two very attractive applications of plural logic: first, as a way of giving an account of sets; second, as a way of obtaining proper classes "for free".

Regrettably, the two applications of plural logic appear incompatible. The first application suggests that any plurality forms a set. Consider any objects *xx*. Presumably, these are what Gödel calls "well-defined objects". If so, it is permissible to apply the "set of" operation to *xx*, which yields the corresponding set {*xx*}. The second application, however, requires that there be pluralities corresponding to proper classes, which by definition are collections too big to form sets. For example, there must be a plurality of all sets whatsoever to serve as the proper class *V*. But, when the "set of" operation is applied to this plurality, we obtain a universal set, which is unacceptable.

Is there any way to retain both of the attractive applications of plural logic? To do so, we would have to restrict the domain of application of the "set of" operation such that the operation is *undefined* on the very large pluralities that correspond to proper classes, while it remains defined on smaller pluralities. The passage from Gödel suggests at least the possibility of such a restriction, because he requires that the "set of" operation be applied to "well-defined objects". The obvious concern is that the needed restriction would be *ad hoc*. The "set of" operation applies to vast infinite pluralities, thus forming large sets in the cumulative hierarchy. But once we allow that these infinite pluralities form sets, why are other infinite pluralities suddenly too large to do so?

7. Towards a reasonable liberalism about definitions

To make progress, we need to take a closer look at our liberal view of definitions. What, exactly, is required for an attempted definition of a set to be permissible?

It is useful to start with an analogy. Suppose you detest web pages that link to themselves.[24] So you wish to create a web page that links to all web pages that are innocent of this bad habit. In other words, you wish to create a web page that links to all and only the web pages that do not link to themselves. Can your wish be fulfilled? The answer depends on how your wish is analyzed. Should the scope of the crucial plural description — 'the web pages that do not link to themselves' — be narrow or wide? Depending on the scope of the description, your wish can be analyzed in either of the following two ways:

(N) You wish to design a web page y such that, for every web page x, y links to x if and only if x does not link to itself.

(W) There are some web pages xx such that, for every web page x, x is one of xx just in case x does not link to itself, and you wish to design a web page y that links to all and only xx.

On the narrow-scope reading (N), your wish is flatly incoherent. The desired web page would have to link to itself just in case it does not link to itself. On this reading, your wish is no better than the wish to bring about the existence of a Russellian barber:

(B) You wish there to be a barber y such that, for all x, y shaves x if and only if x does not shave himself.

On the wide-scope reading (W), by contrast, there is no conceptual or mathematical obstacle to the fulfillment of your wish. First, you identify all the web pages xx that refrain from the bad habit of self-linking. Then, you create a new web page that links to all and only xx.

What explains this stark difference between the two readings? The heart of the matter is how one specifies the target collection — that is, the web pages of which you wish to create a comprehensive inventory. On (N), the target is specified intensionally by means of the condition 'x does not link to itself'. This intensional specification means that the target shifts with the circumstances. First you find that there is no web page of the sort you wish for. So you attempt to fulfill your wish by changing the circumstances, by creating a web page of the desired sort. But since the target is specified intensionally, this new web page must itself be taken into account when assessing whether your wish has been satisfied — which of course it has not, as logic alone informs us.

By contrast, on the wide-scope reading (W), the target is specified extension- ally by means of the plurality xx. This extensional specification ensures that the target stays fixed when you change the circumstances. (Here we invoke the modal rigidity of pluralities, which we defend in [Florio and Linnebo, forthcoming, Chap. 10]). You can thus fulfill your wish by creating a new web page that links to all and only xx. Although xx are described, in

---

[24] This example has been independently used by Brian Rabern in teaching and on social media.

the original circumstances, by means of a condition that is prone to paradox, there is no requirement that *xx* should remain so described in alternative circumstances. Like any other plurality, *xx* are tracked rigidly across alternative circumstances, not in terms of any description that these objects happen to satisfy.

With this analogy in mind, let us return to the question of what is a reasonable liberalism about mathematical definitions. Suppose you care about sets, not web pages. You wish to define a set by specifying its elements. As our webpage analogy reveals, it is essential to distinguish between two different ways in which the elements of the would-be set might be specified. You might specify the elements *intensionally*, by means of a condition φ(x)> :

(I) You wish to define a set y such that, for every object x, x is an element of y if and only if φ(x).

Alternatively, you might specify the elements of the would-be set *extensionally*, by means of a plurality *xx*:

(E) You wish to define a set y such that the elements of y are precisely *xx*. Can either wish be fulfilled?

This is a question about what it takes for a mathematical definition to be permissible. We claim that the proposed definition is often problematic when the target is specified intensionally, but always permissible when the target is specified extensionally. Our defense of these claims will be informed by our web-page analogy.

Let us begin with the negative claim that (I) is often problematic. The reason is simple. We can hardly be more liberal about mathematical definitions than we are about objects that we literally (and easily) construct, such as web pages. This means we need to be extremely cautious about which definitions of sets we deem permissible when the target is specified intensionally. To illustrate how such definitions can be problematic, observe that one instance of the intensionally specified wish (I) is an analogue of the problematic narrow-scope wish (N) concerning web pages:

(N') You wish to define a set y such that, for every object x, x is an element of y if and only if x is not an element of itself.

Just as (N) is flatly incoherent, so, we contend, is (N').

We turn now to our positive claim, namely that the proposed definition is always permissible when the target is specified extensionally by means of a plurality *xx*. Here is the rough idea. The extensional specification ensures that the target will not shift with the circumstances. We therefore have no difficulty making sense of circumstances in which *xx* define a set — much as we have no difficulty making sense of circumstances in which some given web pages *yy* are precisely the ones to which some new web page links.

We can be far more specific, though. Consider a dispute between a proponent and an opponent of the proposed definition. Suppose both parties accept a domain *dd*. The proponent wishes to define one or more sets of the form {*xx*}, where *xx* are drawn from dd. She does not insist that the sets to be defined be among *dd*; in this sense, the sets may be "new". To shore up the proposed definition, she provides the following account of what it takes for "new" sets to be identical or have certain elements:[25]

(i) {*xx*} = {*yy*} if and only if $xx \approx yy$;
(ii) $y \in \{xx\}$ if and only if $y \prec xx$.

These clauses achieve something remarkable. They provide answers to all atomic questions about the "new" sets of the form {*xx*} in terms that are concerned solely with the "old" objects in *dd*, objects that were available before the definition. That is, all atomic questions about the "new" objects receive answers in terms of the "old" objects that both parties to the dispute accept.

Of course, this is merely an instance of the liberal view of definitions that we outlined in Section 5. According to this view, it suffices for a mathematical object to exist that an adequate definition of it can be provided — where the adequacy is understood as follows. Consider a domain *dd* of objects standing in certain relations. We would like to define one or more additional objects. Suppose our definition provides truth conditions for any atomic predication concerned with the desired "new" objects in the form of some statement concerned solely with the "old" objects with which we began. Thus, any atomic question about the "new" objects can be reduced to a question that is solely about the "old" objects. Then, according to our liberal view, the definition is permissible.

It is instructive to compare with the situation where the desired set is specified intensionally, by means of a membership condition for each of the desired sets. Again, we start with some objects *dd* accepted by both parties. A more extreme proponent of liberal definitions may wish to define sets of the form $\{x : \varphi(x)\}$, where any parameters in the membership condition $\varphi(x)$ are drawn from *dd*. Again, she does not insist that these sets be among dd; they may be "new". The opponent will rightly challenge her to provide an account of what it takes for "new" sets to be identical or to have certain elements. Given the intensional specification of the desired sets, her answers will be as follows:

(i') $\{x : \varphi(x)\} = \{x : \psi(x)\}$ if and only if $\forall x(\varphi(x) \leftrightarrow \psi(x))$;
(ii') $y \in \{x : \varphi(x)\}$ if and only if $\varphi(y)$.

---

[25] If *dd* already contain some sets, then clause (i) must be understood to range over both "old" and "new" sets.

These answers are potentially problematic in a way that their extensional analogues, (i) and (ii), are not. An interesting example is the attempt to define a set $a = \{x : x \in x\}$. If this definition is to succeed, there must be an answer to the question of whether a is an element of itself. But the only answer we receive from clause (ii′) is that $a \in a$ if and only if $a \in a$. Of course, this is useless.[26] More tellingly, the answer is not stated in terms of the objects accepted by both parties to the dispute. An atomic question about the "new" object a receives an answer that essentially involves *this very object*; there is no reduction to the "old" objects among dd.

Notice that it is of no avail for the extreme liberal to allow a to lie outside of dd, that is, in our parlance, to be "new". The set a is specified intensionally, by means of the membership condition '$x \in x$', and we cannot "outrun" this specification. Even in a domain that strictly extends dd, a is, by definition, the set of all and only the objects that satisfy the condition '$x \in x$'. By contrast, when a set is specified extensionally by means of a plurality xx, it does help to consider a domain that strictly extends dd. Even if xx are, say, all the sets among dd that are not elements of themselves, xx need not satisfy this plural description in an extended domain; for xx are tracked rigidly into the extended domain, not by means of the description. This makes the world safe for the desired set {xx}, provided that the set is located outside of dd. Notice also the striking parallelism with the case of web-page design. Suppose you want a web page to link to all and only the members of some collection of web pages, for example, the collection of web pages that do not link to themselves. If the target collection is specified intensionally, it is of no avail to create *a new* web page: you cannot "outrun" this problematic specification. By contrast, if the collection is specified extensionally, there is no obstacle to the creation of the desired web page.

The picture that emerges is that there is a fundamental difference between the proposed definitions of sets depending on whether the target is specified extensionally or intensionally. In the former case, every atomic question about the "new" objects receives an answer expressed solely in terms of the "old" objects, whereas in the latter case, this kind of reduction is often unavailable. The proposed definitions are therefore often unacceptable when the target is specified intensionally. In the case of an extensional specification, on the other hand, a proponent of liberal definitions is in a much stronger position. She has laid out certain definitions, which are mathematically fruitful and have the desirable property that all atomic questions about the "new" objects receive answers in terms that are acceptable to her opponent.

---

[26] It could be worse. When we ask whether the Russell set $b = \{x : x \notin x\}$ is an element of itself, we receive an inconsistent answer.

Admittedly, the proponent of liberal definitions cannot force her opponent to accept the proposed definitions: he does not contradict himself when he rejects them. But she can justifiably accuse her opponent of dogmatism that stifles scientific progress. He dogmatically clings to certain beliefs in a way that stands in the way of fruitful mathematics. By insisting that *dd* are allencompassing — and thus that there can be no "new" objects outside of *dd* — he privileges certain metaphysical or logical dogmas over good mathematics. We can hardly think of a better way to defend this outlook than by quoting the following passage by Cantor.

> Mathematics is in its development entirely free and only bound in the self-evident respect that its concepts must both be consistent with each other and also stand in exact relationships, ordered by definitions, to those concepts which have previously been introduced and are already at hand and established. [...] [T]he essence of mathematics lies precisely in its freedom.([Cantor, 1883, pp. 19–29], as translated in [Ewald, 1996, p. 896])

8. The principles of critical plural logic

We have argued that any given objects can be used to define a set. Unsurprisingly, this has consequences for our choice of a plural logic. To avoid paradox, we have little choice but to restrict the plural comprehension scheme. Contrary to what has traditionally been assumed, not every condition defines a plurality. To emphasize this departure from traditional plural logic, let us call our approach *critical plural logic*.[27] The acceptance of this approach has implications far beyond the philosophy of mathematics, affecting views in semantics and metaphysics that rely on traditional plural logic. We provide a detailed assessment of our approach *vis-à-vis* the traditional one in [Florio and Linnebo, forthcoming, esp. Chaps 2, 11].

How, exactly, does our critical plural logic differ from the traditional version? We accept the usual sentential and first-order logic. Furthermore, we allow the plural quantifiers to be governed by axioms and rules analogous to those governing the first-order quantifiers.[28] Our quarrel with traditional plural logic concerns only the question of what pluralities there are, or, in other words, the question of which plural comprehension axioms to accept. It is therefore incumbent on us to clarify what pluralities we take there to be. It is insufficient merely to observe that the plural comprehension scheme needs to be restricted in some way or other to avoid a universal plurality. We need some "successor principles" to the unrestricted plural comprehension scheme that tell us what pluralities there in fact are.

---

[27] We motivate this label in Section 11.
[28] But, of course, we should insist that the formulation of logical rules be neutral with respect to which comprehension axioms are validated.

How should these successor principles be chosen and motivated? When discussing this question, we believe it is useful to keep in mind the following intuitive version of our argument for restricting the plural comprehension scheme.

> To define a plurality, we need to circumscribe some objects. When we circumscribe some objects, however, we can use these objects to define yet another object, namely their set. Since yet another object can in this way be defined, it follows that the circumscribed objects cannot have included *all* objects. Thus, reality as a whole cannot be circumscribed: there is no universal plurality. Consequently, the plural comprehension scheme needs to be restricted.

This argument hinges on the idea that every plurality is circumscribed, or, as we will also put it, *extensionally definite*.

Can this notion of extensional definiteness guide our search for successor principles and justify, or at least motivate, the resulting principles? Here we face a fork in the road, depending on whether or not we attempt to provide an analysis of extensional definiteness in more basic terms, and on this basis, to provide the requisite guidance and justification.

There have been several attempts to provide such an analysis. Linnebo [2013] proposes a modal analysis inspired by Cantor's famous distinction between "consistent" and "inconsistent" multiplicities. Here is how Cantor explains the distinction in a famous letter to Dedekind of 1899:

> [I]t is necessary... to distinguish two kinds of multiplicities (by this I always mean definite multiplicities). For a multiplicity can be such that the assumption that all of its elements 'are together' leads to a contradiction, so that it is impossible to conceive of the multiplicity as a unity, as 'one finished thing'. Such multiplicities I call *absolutely infinite or inconsistent multiplicities*... If on the other hand the totality of the elements of a multiplicity can be thought of without contradiction as 'being together', so that they can be gathered together into 'one thing', I call it a *consistent multiplicity* or a 'set'. (As quoted in [Ewald, 1996, pp. 931–932])

Using the resources of modal logic, it is relatively straightforward to formalize Cantor's notion of a multiplicity being "one finished thing", namely that it is possible for all possible members of the multiplicity to exist or "be together". Or, changing the idiom slightly, there is no possibility of the multiplicity gaining yet more members at more populous possible worlds. Based on this analysis, Linnebo [2013] proves various principles of extensional definiteness, which in the present context amount to principles concerning the existence of pluralities.

Another analysis of extensional definiteness is inspired by Michael Dummett's suggestion that a domain is definite just in case quantification over this domain obeys the laws of classical logic, not just intuitionistic.[29] Intrigu-

---

[29] A closely related idea is found in Solomon Feferman's widely circulated and discussed manuscript, "The continuum hypothesis is neither a definite mathematical problem nor a definite

ingly, it turns out that a fairly natural development of this Dummettian suggestion validates almost the same principles of extensional definiteness as the modal analysis [Linnebo, 2018]. Yet other analyses may be possible as well. We invite the readers to explore.

Here we wish to pursue the other fork in the road, namely to leave the notion of extensional definiteness unanalyzed and instead to use our intuitive conception of the notion, coupled with abductive considerations, to motivate principles of extensional definiteness. This strategy has both advantages and disadvantages: it is more general, as it avoids specific theoretical commitments; but it also provides less leverage and thus less of an independent check on the proposed principles of definiteness. In any case, we believe this is an option worth exploring. We thus ask what it is for a collection to be circumscribed or extensionally definite.

First, since every single object can be circumscribed, there are singleton pluralities:

$$\forall x \exists yy \forall z (z \prec yy \leftrightarrow z = x).$$

Second, because the result of adding one object to a circumscribed plurality is also circumscribed, we accept a principle of adjunction. Given any plurality xx and any object y, we can adjoin y to xx to form the plurality $xx + y$ defined by:

$$\forall u(u \prec xx + y \leftrightarrow u \prec xx \vee u = y).$$

Moreover, a plural separation principle is well motivated. Suppose you have circumscribed a collection and have formulated a sharp distinction between two ways that members of the collection can be. Then the subcollection whose members are all and only the objects that lie on one side of this distinction is in turn circumscribed. More formally, given any plurality xx and any condition $\varphi(x)$ that has an instance among xx, there is a plurality yy of those members of xx that satisfy the condition:

$$\exists x(\varphi(x) \wedge x \prec xx) \to \exists yy \forall u(u \prec yy \leftrightarrow u \prec xx \wedge \varphi(u)).$$

Next, there are some plausible union principles. Let us begin with a simple case. Since two circumscribed collections can be conjoined to make a single such collection, a principle of pairwise union is plausible. Given any plurality xx and any objects yy, there is a union plurality zz defined by:

$$\forall xx \, \forall yy \, \exists zz \, \forall u(u \prec zz \leftrightarrow u \prec xx \vee u \prec yy).$$

---

logical problem" [Feferman, unpublished].

A generalized union principle can also be motivated. Consider some circumscribed collections, each with its own unique tag. Suppose that the collection of tags is also circumscribed. Then the "union collection" comprising all the items that figure in at least one of the tagged collections is circumscribed. This motivates a generalized union principle to the effect that the union of an extensionally definite collection of extensionally definite collections is itself extensionally definite. We can formulate this as the following schema. Suppose there are *xx* such that:

$$\forall x(x \prec xx \to \exists yy \, \forall z(z \prec yy \leftrightarrow \psi(x,z))).$$

Then there are *zz* such that:

$$\forall y(y \prec zz \leftrightarrow \exists x(x \prec xx \land \psi(x,y))).$$

Although the generalized union principle does not, on its own, entail the pairwise one, this entailment does go through in the presence of the singleton and adjunction principles.[30] It therefore suffices to adopt the generalized union principle.

The principles accepted so far do not entail the existence of any infinite pluralities; indeed, they have a model where every plurality is finite. Is it possible for an infinite collection to be circumscribed and thus to correspond to a plurality? This question calls to mind the ancient debate about the existence of completed infinities. Aristotle famously argued that only finite collections can be circumscribed, and that a collection can be infinite only in the potential sense that there is no finite bound on how many members the collection might have. This remained the dominant view until Cantor, who boldly defended the actual infinite and the existence of completed infinite collections. The natural numbers provide an example. Aristotle denied, whereas Cantor affirmed, the existence of a completed collection of all natural numbers.

We are interested in an analogous question concerning pluralities. Let '$P(x, y)$' mean that x immediately precedes y. Following first-order arithmetic, we accept that every natural number immediately precedes another:[30][31]

$$\forall x \, \exists y P(x,y). \quad (1)$$

---

[30] Proof sketch. Consider two pluralities *xx* and *yy*. Assume there are two distinct objects, say a and b, to tag these pluralities. (If there is only a single object, the pairwise union of *xx* and *yy* is a singleton plurality.) Now apply the generalized union principle to the formula '$(x = a \land y \prec xx) \lor (x = b \land y \prec yy)$', observing that a and b form a plurality. This yields the pairwise union of *xx* and *yy*.

[31] Aristotle would only accept a weaker, modal analogue of this principle, namely $\Box \, \forall x \Diamond \, \exists y P(x,y)$, where the modal operators represent metaphysical modalities.

We would like to know whether there is a circumscribed collection, or plurality, of all natural numbers. More precisely, we would like to know whether there are some objects xx containing 0 and closed under P, in the following sense:

$$\exists xx(0 \prec xx \land \forall x \, \forall y(x \prec xx \land P(x,y) \to y \prec xx)). \quad (2)$$

Although asserting the existence of such a plurality is a substantial step, it has also been a tremendous theoretical success, as mathematics since Cantor has made amply clear. On abductive grounds, we therefore recommend accepting (2), conditional on (1), as a plural analogue of the set-theoretic axiom of Infinity.

It will be objected that this conditional principle is concerned specifically with the natural numbers and thus lacks the topic neutrality of a logical law. The objection is entirely reasonable and points to the need for a more general principle that justifies transitions such as the one from (1) to (2). There is nothing special about 0 and the functional relation P. So, for any plurality xx and functional relation, there should be a plurality yy containing xx and closed under that function. We therefore claim that the desired generalization is the schematic principle that every plurality can be closed under function application:

$$\exists xx(0 \prec xx \land \forall x \, \forall y(x \prec xx \land P(x,y) \to y \prec xx)). \quad (3)$$

We adopt this as the official plural principle of infinity. In practice, however, it does not much matter whether we accept this more general schematic principle or merely (2), conditional on (1). For in the presence of first-order arithmetic, ordered pairs, and the other principles concerning pluralities, these two principles of infinity are provably equivalent.[32]

A plural analogue of the axiom of Replacement is plausible as well. Consider a plurality of objects. Now you may replace any member of this plurality with any other object, or, if you prefer, leave the original object unchanged. Then the resulting collection is also circumscribed and thus defines a plurality of objects. We formalize this as follows.

---

[32] Proof sketch. The only hard direction is to show that the specific conditional entails the general one. Consider any xx, and assume that ψ is functional. For every member $a \prec xx$, we contend that there is a plurality zz_a containing a and closed under ψ. Given this contention, the generalized union principle enables us to define the desired plurality yy as the union of all the pluralities zz_a. To prove the contention, we observe that, using ordered pairs and plural quantification, we can produce a formula $\theta(n,y)$ which expresses that n is a natural number and that y is the nth successor of a in the series generated by ψ. We do this by letting $\theta(n,y)$ state that $\langle n, y \rangle$ is a member of every plurality containing $\langle 0, a \rangle$ and closed under the operation $\langle m, u \rangle \mapsto \langle m+1, v \rangle$, where v is the unique object such that $\psi(u,v)$. Now we apply the generalized union principle to the plurality of all natural numbers and the formula θ to obtain the desired plurality zz_a.

$$\forall xx[\,\forall x(x \prec xx \to \exists! y \psi(x,y)) \to \exists yy \forall y(y \prec yy \leftrightarrow \exists x(x \prec xx \wedge \psi(x,y)))].$$

It is pleasing to observe that this plural version of Replacement follows from the generalized union principle and the singleton principle. And, as in the case of sets, the plural principle of replacement entails that of separation.[33]

To sum up, our intuitive conception of extensional definiteness motivates the following three principles concerning pluralities:

- singleton,
- adjunction,
- generalized union.

An additional principle receives a more theoretical justification:

- infinity.

These four principles constitute the system we call critical plural logic.

As observed, the first three of these principles entail some other plausible principles:

- separation,
- pairwise union,
- replacement.

Moreover, it is straightforward to verify that each principle of critical plural logic can be derived from traditional plural logic. In essence, each of the pluralities we licence is a subplurality of the universal plurality licenced by traditional plural logic. Critical plural logic is therefore strictly weaker than the traditional system. This relative weakness is for a good cause, as will emerge clearly in Section 10, where we explore the connection between critical plural logic and set theory. This connection is far simpler and, we believe, more natural, than in the case of traditional plural logic.

9. Extensions of critical plural logic

When stronger expressive resources are accepted, various extensions of critical plural logic can be formulated and justified. Suppose there are "superpluralities", that is, pluralities of pluralities. As customary, we let triple variables, such as 'xxx', have superplural reference. This addition enables

---

[33] *Proof sketch.* Consider xx and a condition $\varphi(x)$. Assume $\varphi(a)$ for some member a of xx. Now apply the principle of replacement to the condition $\psi(x,y)$ defined as $(\neg\varphi(x) \wedge y = a) \vee (\varphi(x) \wedge y = x)$. This yields the subplurality of those members of xx that satisfy $\varphi(x)$.

us to express superplural analogues of the principles of critical plural logic. Here we will focus on two more interesting, novel principles.

First, we can formulate a principle of extensional definiteness that corresponds to the familiar set-theoretic axiom of Powerset. We can do this entirely without mention of sets by using superplurals. For any plurality xx, there is a superplurality yyy of all subpluralities of xx:

$$\forall xx \, \exists yyy \, \forall zz(zz \prec yyy \leftrightarrow zz \preceq xx). \qquad (4)$$

The justification for this "powerplurality" principle is less straightforward than in the case of the earlier principles. It relies on what Bernays (1935) calls "quasi-combinatorial" reasoning: a combinatorial principle that is compelling for finite domains is extrapolated to infinite domains. The powerplurality principle is certainly reasonable when the plurality xx is finite: we can then list all of its subpluralities. The general principle is a big and admittedly daring extrapolation of the finitary principle into the infinite. Its justification is partially abductive. Just like its set-theoretic analogue, the principle fits into a coherent and fruitful body of theory, as will be explained shortly. The principle also provides important information about which superpluralities there are.

Second, superplurals make it possible to formulate plural choice principles. For example, given a superplurality xxx of non-overlapping pluralities, there is a "choice plurality" whose members include one member of each plurality of xxx. That is, for each such xxx we have:

$$\exists yy \, \forall zz(zz \prec xxx \rightarrow \exists! y(y \prec zz \wedge y \prec yy)).$$

As in the case of the powerplurality principle, plural choice principles are extrapolations from the finite into the infinite, and their is justification is partially abductive.[34]

In sum, the addition of superplural resources enables us to formulate and justify an extended critical plural logic. Two distinctive principles are:

- powerplurality
- choice

Of course, yet stronger principles can be countenanced as ever greater expressive resources are considered.

---

[34] See [Pollard, 1988] for a defense of the Axiom of Choice on the basis of a plural choice principle. If ordered pairs are available, there is less of a need for superplurals to express choice principles. For example, we can assert that for any relation coded by means of a plurality of ordered pairs, there is a functional subrelation with the same domain, again coded by means of a plurality of ordered pairs.

10. Critical plural logic and set theory

The various plural principles we have discussed provide valuable information about sets. To see this, recall the correspondence we have advocated between pluralities and sets:

(i) $\{xx\} = \{yy\}$ if and only if $xx \approx yy$;
(ii) $y \in \{xx\}$ if and only if $y \prec xx$.

Using this correspondence, the plural principles entail analogous set-theoretic axioms.

However, there are two reasons to worry that the plural principles will not lead to ordinary ZFC. First, since we do not ordinarily admit an empty plurality, there is a threat of losing the empty set. Some ways to address this threat were discussed in Section 5. One solution is to allow an empty plurality. Another is to allow the set-of operator $xx \mapsto \{xx\}$ to be what Oliver and Smiley [2016, p. 88] call a "co-partial" function, which can thus take the value $\emptyset$ on an undefined argument. Either way, we can prove the existence of an empty set.

Second, since plural logic is applied to all sorts of objects, the mentioned correspondence introduces impure sets, that is, sets of non-sets. The relevant comparison is therefore not ZFC, but ZFCU — a modified system which accommodates urelements. This system is obtained by making explicit the quantification over sets in the axioms of ZFC. Whenever a quantifier of an axiom of ZFC is intended to range over sets even when urelements are introduced, we explicitly restrict this quantifier to sets by means of a predicate 'S' intended to be true of all and only sets. For example, the axiom of Extensionality is rewritten as:

$$\forall x \forall y [S(x) \wedge S(y) \to (\forall u(u \in x \leftrightarrow u \in y) \to x = y)].$$

Our aim, then, is to use critical plural logic and the correspondence principles (i) and (ii) to derive axioms of ZFCU. We define '$S(x)$' as '$\exists xx(x = \{xx\})$'. This enables us to derive the axioms of Empty Set, Pairing, Separation, Union, Infinity, and Replacement. Moreover, the axiom of Extensionality follows immediately from the correspondence between pluralities and sets, and Foundation too can be seen as explicating the way in which sets are successively formed from pluralities of elements.

To derive the axioms of Powerset and Choice, we need to go beyond critical plural logic. Choice follows naturally from the superplural choice principle discussed in the previous section. Deriving Powerset is less straightforward. Given any set a, we want to prove the existence of its powerset. To do so, we need to show that there is a plurality of all of a's subsets. How might this be done? One option, inspired by the iterative conception of set, is to postulate the existence of such a plurality, on the grounds that when a was formed, all its elements were available, thus giving us the ability also to form all of a's subsets. We prefer to utilize the powerplurality principle of the previous section, reasoning as follows. Let aa be the elements of a, and consider their superplurality bbb. For every subset x of a, if x = {xx} for some xx, then xx ≺ bbb. That is, bbb circumscribe all the subpluralities of aa. But if some pluralities are jointly circumscribed, so are the unique sets formed from precisely these pluralities. This gives us the desired plurality of subsets of a. (This reasoning assumes that the extended, superplural logic contains a replacement principle that allows us to replace each plurality of a superplurality with a unique object and thus arrive at a plurality.)

Our discussion shows that critical plural logic, and the plausible superplural extensions thereof, have great explanatory power, especially in connection with the correspondence principles (i) and (ii). Still, one might worry that things are too good to be true. Do we even know that our assumptions — the mentioned plural logics and the correspondence principles — are jointly consistent? This worry can be put to rest by proving that these assumptions are consistent relative to ZFC. For critical plural logic and the correspondence principles, we do this by translating plural quantifiers as first-order quantifiers restricted to non-empty sets. An analogous relative consistency result can be given for the described extension of critical plural logic. In that case, superplural quantifiers are translated as first-order quantifiers restricted to non-empty sets of non-empty sets.

On the view we have defended, plural logic lacks one of the features commonly ascribed to pure logic, namely epistemic primacy vis- à-vis all other sciences (see Section 2). To see this, we need only recall the extent to which our defense of critical plural logic relies on abductive considerations, in particular, on considerations about what constitutes a permissible mathematical definition. Moreover, some of the principles of critical plural logic—infinity, powerplurality, and choice—specifically received an abductive justification.

Our view forges a close connection between the principles of critical plural logic and the axioms of set theory, which suggests that critical plural logic and its extensions have non-trivial mathematical content. Let us explain. We have provided a factorization of set theory into two components: the correspondence principles, which link pluralities and their corresponding sets, and critical plural logic, which provide information about what pluralities there are and how these behave. Clearly, the strong mathematical content of set theory derives from these two components. It is the correspondence principles that introduce sets as mathematical objects by characterizing what Gödel called the "the set of" operation" (see Section 5). What sets there are, however, will depend on what pluralities are "fed into" this operation and is determined in large part by the plural logic that is brought to bear. We can study this dependence by keeping the correspondence principles fixed, while varying the plural logic to which they are applied. As just observed, our extended critical plural gives rise to full Zermelo-Fraenkel set theory. If we remove the plural principle of infinity, the result is a comparatively weak theory of hereditarily finite set. Alternatively, suppose we retain that plural principle of infinity but impose a predicativity requirement on the generalized union principle (and thus also plural replacement and separation).[35] Then a broadly predicative set theory ensues. In short, when we keep the correspondence principles fixed but vary the plural logic, we obtain set theories with wildly different mathematical content. This observation strongly suggests that some of the mathematical content of the resulting set theory derives from the plural logic to which the correspondence principles are applied, not solely from these principles. If correct, it follows that a theory can have substantial mathematical content without any commitment to mathematical objects.

To come to terms with the possibility of mathematical content even in the absence of mathematical objects, it is useful to recall Bernays's notion of quasi-combinatorial reasoning, whereby principles that are compelling in finite domains are extrapolated to infinite ones. Bernays and others regard such reasoning as distinctively mathematical and a major watershed in the foundations of mathematics, marking the onset of serious infinitary reasoning. Since critical plural logic and its extensions embody, and are motivated by, such reasoning, Bernays would regard both the notion of a plurality and the principles of critical plural logic as distinctively mathematical in character. This is particularly clear for the plural principles of infinity, powerplurality, and choice, whose justification explicitly relied on quasi-combinatorial reasoning.

Is the mathematical content of plural logic compatible with our view that pluralities can be used to explain sets? We believe it is. The explanation in question is a broadly metaphysical one: we make sense of a set {xx} as "formed" from its elements xx. There is no conflict between this explanation and the view that plural logic has non-trivial mathematical content. Indeed, on this view, the indisputable mathematical content of set theory is in part inherited from that of plural logic.[36]

11. Concluding remarks

Two larger questions have pervaded our entire discussion. The first of these questions concerns how we choose a "correct" logic. Some starkly different views are found in the literature. At one extreme we find Frege, who claims that logic codifies "the basic laws" of all rational thought, and the laws of logic must therefore be presupposed by all other sciences. He writes:

---

[35] Specifically, we require that the formula $\psi(x, y)$ be predicative, in the sense that it contain no bound plural variables.
[36] Thanks to Hans Robin Solberg for raising this concern.

I take it to be a sure sign of error should logic have to rely on metaphysics and psychology, sciences which themselves require logical principles. [Frege, 1903, p. xix]

This "logic first" view has been very influential. Following Frege, logic is often regarded as epistemologically and methodologically fundamental. All disciplines, including mathematics, are answerable to logic rather than vice versa.

At the opposite extreme we find Quine, whose radical holism leads him to assimilate logic and mathematics to the theoretical parts of empirical science. Logic and mathematics, he claims, are not essentially different from theoretical physics: although they go beyond what can be observed by means of our unaided senses, they are justified by their contribution to the prediction and explanation of states of affairs that *can* be observed.

These extremes are not the only views, however. In particular, one need not be a radical holist to reject the Fregean logic-first view. What are sometimes called "critical views of logic" represent a less dramatic departure from Frege.[37] These views hold that the logical principles governing some subject matter may depend on features of this subject matter or of our discourse about it. The views thus stop short of Quine's radical holism and emphasize instead a more local entanglement of logic with some particular discipline, such as mathematics, semantics, or some part of metaphysics. As a result of this entanglement, logic is answerable to one's views in this other discipline.

The revision of plural logic that we have defended provides a good example of such a critical view of logic. Avoiding any commitment to Quinean holism, we have argued that the principles of plural logic are entangled with our theory of correct mathematical definitions. Specifically, we have defended a liberal theory of mathematical definitions, and on the basis of that theory, we have argued that plural comprehension needs to be restricted more than has traditionally been assumed.

The second larger question on which this article bears concerns the relation between the philosophy of mathematics and mathematical practice. It is obviously a good thing when the philosophy of mathematics is informed by mathematical practice — just as the philosophy of *any* special science ought to be informed by the practice of that science. We do not regard this approach to scientific practice as a threat to traditional philosophy of mathematics. The reason is simple. As we have argued, mathematical practice does not always speak with a single voice. In cases where it does not, there is no easy way to extract philosophical or methodological lessons from mathematical practice. Thus, even for those of us who want our

---

[37] See [Parsons, 2015] and a forthcoming special issue of Inquiry edited by Mirja Hartimo, Frode Kjosavik, and Øystein Linnebo.

philosophy of mathematics to pay close attention to mathematical practice, there remains an important role for more traditional forms of philosophical analysis.

References

Benacerraf, P., and H. Putnam, eds [1983]: Philosophy of Mathematics: Selected Readings. 2nd ed. Cambridge University Press.
Black, M. [1971]: 'The elusiveness of sets', *Review of Metaphysics* **24**, 614–636.
Boolos, G. [1984]: 'To be is to be a value of a variable (or to be some values of some variables)', *Journal of Philosophy* **81**, 430–449.
_____ [1985]: 'Nominalist platonism', *Philosophical Review* **94**, 327–344.
Burgess, J., and G. Rosen [1997]: *A Subject with No Object. Strategies for Nominalistic Interpretations of Mathematics*. Oxford University Press.
Cantor, G. [1883]: *Grund lag en einer allgemeinen Mannichfaltigkeitslehre. Ein mathematisch-philosophischer Versuch in der Lehre des Unend lichen*. Leipzig: Teubner.
_____ [1895]: 'Beitrage zur BegrUndung der transfiniten Mengenlehre I', *Mathematische Annalen* **46**, 481–512.
Ewald, W., ed. [1996]: *From Kant to Hilbert: A Source Book in the Foundations of Mathematics*. Vol. 2. Oxford University Press.
Feferman, S. [unpublished]: 'The Continuum Hypothesis is neither a definite mathematical problem nor a definite logical problem'. Available at https://math.stanford.edu/~feferman/papers/CH is Indefinite.pdf. Accessed May 2020.
Fine, K. [1991]: 'The study of ontology', *Noûs* **25**, 263–294.
Florio, S., and Ø. Linnebo [2016]: 'On the innocence and determinacy of plural quantification', *Noûs* **50**, 565–583.
_____ [forthcoming]: The *Many and the One: A Philosophical Study*. Oxford University Press.
Frege, G. [1893/1903]: *Grundgesetze der Arithmetik*. Band I/II. Jena: Verlag Hermann Pohle. Eng. title: *Basic Laws of Arithmetic*. P.A. Ebert and M. Rossberg, trans. Oxford University Press, 2013.
Gödel, K. [1964]: 'What is Cantor's continuum problem?', in S. Feferman, *et al.*, eds, *Collected Works, Volume II, Publications 1938–1974*, pp. 176–188. Oxford University Press, 1990.
Hewitt, S. [2015]: 'When do some things form a set?', *Philosophia Mathematica* (3) **23**, 311–337.
Higginbotham, J. [1998]: 'On higher-order logic and natural language', in T. Smiley, ed., *Philosophical Logic*, pp. 1–27. Oxford University Press.
Hossack, K. [2000]: 'Plurals and complexes', *British Journal for the Philosophy of Science* **51**, 411–443.
Kreisel, G. [1967]: 'Informal rigour and completeness proofs', in I. Lakatos, ed., *Problems in the Philosophy of Mathematics*, pp. 38–186. Amsterdam: North-Holland.
Lewis, D. [1991]: *Parts of Classes*. Oxford: Blackwell.
Linnebo, Ø. [2003]: 'Plural quantification exposed', *No^s* **37**, 71–92.
_____ [2013]: 'The potential hierarchy of sets', *Review of Symbolic Logic* **6**, 205–228.
_____ [2018]: *Thin Objects: An Abstractionist Account*. Oxford University Press. Maddy, P. [1983]: 'Proper classes', *Journal of Symbolic Logic* **48**, 113–139.
McKay, T.J. [2006]: *Plural Predication*. Oxford University Press.
Oliver, A., and T. Smiley [2001]: 'Strategies for a logic of plurals', *Philosophical Quarterly* **51**, 289–306.

_____ [2016]: *Plural Logic*. 2nd ed. Oxford University Press.
Parsons, C. [1977]: 'What is the iterative conception of set?', in R.E. Butts and J. Hintikka, eds, *Logic, Foundations of Mathematics, and Computability Theory*, pp. 335–367. Dordrecht: Reidel. Reprinted in [Benacerraf and Putnam, 1983] and [Parsons, 1983].
_____ [1983]: *Mathematics in Philosophy*. Ithaca, N.Y.: Cornell University Press.
_____ [2015]: 'Infinity and a critical view of logic', *Inquiry* **58**, 1–19.
Pollard, S. [1988]: 'Plural quantification and the axiom of choice', *Philosophical Studies* **54**, 393–397.
Quine, W.V. [1982]: *Methods of Logic*. 4th ed. Harvard University Press.
Rayo, A. [2002]: 'Word and objects', *Nous* **36**, 436–464.
Reinhardt, W. [1974]: 'Set existence principles of Shoenfield, Ackermann and Powell', *Fundamenta Mathematicae* **84**, 5–34.
Resnik, M. [1988]: 'Second-order logic still wild', *Journal of Philosophy* **85**, 75–87.
Russell, B. [1903]: The Principles of Mathematics. 2nd ed. Cambridge University Press.
Schein, B. [1993]: Plurals and Events. Cambridge, Mass.: MIT Press.
Simons, P. [2016]: 'The ontology and logic of higher-order multitudes', in A. Arapinis, M. Carrara and F. Moltmann, eds, Unity and Plurality: Logic, Philosophy, and *Linguistics*, pp. 55–89. Oxford University Press.
Yi, B.-U. [1999]: 'Is two a property?' *Journal of Philosophy* **96**, 163–190.
_____ [2002]: *Understanding the Many*. London: Routledge.
_____ [2005]: 'The logic and meaning of plurals. Part I', *Journal of Philosophical Logic* **34**, 459–506.
_____ [2006]: 'The logic and meaning of plurals. Part II', *Journal of Philosophical Logic* **35**, 239–288.

Kant on the possibilities of mathematics and the scope and limits of logic

Frode Kjosavik

ABSTRACT
I suggest how a broadly Kantian critique of classical logic might spring from reflections on constructibility conditions. According to Kant, mathematics is concerned with objects that are given through 'arbitrary synthesis,' in the form of 'constructions of concepts' in the medium of 'pure intuition.' Logic, by contrast, is narrowly constrained – it has no objects of its own and is fixed by the very forms of thought. That is why there is not much room for developments within logic, as compared to the progress in mathematics. Kant's view of logic remains critical, though – through considerations that are effectively on the scope and limits of classical logic and which play a part in his transcendental idealism. The most important ones are to be found in his critique of the use of *reductio ad absurdum* proofs in metaphysics and his solutions to the 'antinomies of pure reason.' Arguably, these considerations carry over to mathematics as well – by way of 'analogues' to the antinomies – in particular in resolving infinity paradoxes.

1. Introduction: The Critical View

Kant's position in philosophy was critical – not 'dogmatic' or 'skeptical.' Still, his view of logic is often considered to be uncritical.[1] After all, in the *Critique of Pure Reason* logic is said to be essentially unrevised since Aristotle, not having taken a single step 'backwards' or 'forward,' and it

---

[1]Gila Sher argues that Kant accords an *exceptional* status to logic:

> … Kant's approach to logic is quite different from his approach to other disciplines. While Kant set out to provide a substantive foundation for scientific and mathematical knowledge, he took formal logic largely as given. Kant's "Copernican Revolution" did not include a revolution in logic … ". (Sher 2016, 241–242)

Cf. also Mirja Hartimo, in the present issue, on Husserl's criticism of Kant's 'uncritical' attitude to logic. It is still possible to hold that Kant's transcendental idealism extends to logic as well, as in Maddy (2007, 223), according to which logic 'depends only on the discursive features of our understanding, not on our particular forms of intuition.' It is therefore 'weakly transcendentally ideal.'

appears, therefore, to be 'finished and complete' (CPR, Bviii). Despite this, Kant did in fact subject logic to a critique of sorts in the Antinomy of Pure Reason. As it turns out, this is of relevance to the philosophy of mathematics. In the following, I shall present a *critical view* of logic that is shaped by distinctly Kantian concerns:[2]

There can be a basis in mathematics for *limitations in a logical principle's applicability*, in particular that of the Law of Excluded Middle (LEM).

It contrasts with a *priority view* of classical logic, according to which its principles are universally valid and applicable to any discourse. On that view, LEM holds and a mathematical statement is either true or false. A statement's truth does not hinge on our having a way of proving it. In accordance with the correspondence notion of truth, mathematical statements describe determinate mathematical objects that exist and have properties independently of whether we have proof access to them or not – as in mathematical Platonism.

Intuitionists within the philosophy of mathematics, on the other hand, reject the universal validity of LEM. They insist on proofs for there to be truth or falsehood and demand that the proofs be constructive. They deny that there is a robust, mind-independent reality that true mathematical statements are true of, or at least they deny that we can make any use of an assumption of realism. Their view of logic is not critical in the moderate sense specified above. Rather, it is the polar opposite of the priority view of logic, namely, *a priority view of mathematics*. It is mathematical activity – when rendered sufficiently constructive – that is closely interwoven with human thought. The subject matter of logic, on the other hand, is confined to regularities in the descriptions of this activity.

Kant might be said to be an 'intuitionist' and a 'constructivist' in the following sense. He demands that all mathematical *concepts* – those with referents as well as mere operational ones – be constructible, i.e. that they can be exhibited in the medium of pure intuition. The constructions are of two kinds – either *ostensive* or *symbolic* (cf. CPR A717/B745; A734/B762). In ostensive constructions, 'objects' that fall under or 'correspond to' mathematical concepts are exhibited in intuition. Only geometrical figures – or the concepts thereof – are explicitly said to be constructible in this way in Kant. But it can reasonably be taken to cover other cases

---

[2] A 'critical view of logic' is discussed by Parsons (2015), with a focus on Brouwer, Weyl and Hilbert, as well as on the 'entanglement' of logic and mathematics. It is not analyzed in relation to Kant's own view of logic, though, but the label is intended to be Kantian in spirit. Notably, in 'Kant's Philosophy of Arithmetic,' Parsons makes the following suggestive remark on Kant's own *uncritical* view of logic: 'But what would have been needed for Kant to be dissatisfied with "traditional logic" might only have been more insight into his own discoveries.' (1983b, 116.)

as well, in particular finite sets as well as strings of strokes for numbers. There are certainly passages that can be construed in that way, like the remark on Segner's arithmetic, in which the concept of the number 5 is said to 'correspond to' five visible 'points' (B15). In symbolic constructions, on the other hand, there are mere symbols – mathematical objects are not exhibited. Still, structural transparency is achieved by way of the intuitable notation they rely on. Symbols are placed together 'in accordance with the form of notation' for various operations (A717/B745), and what is achieved thereby is something 'which discursive cognition could never achieve by means of mere concepts.' (A717/B745). Thus, what is intuitable in mathematics for Kant comprises not only objects but also executions of operations, and not only translations and rotations in geometry, or the finite iteration of these operations, but also step-by-step counting and calculation in arithmetic, or symbol manipulation as laid down by rules in algebra.

However, unlike Brouwer, Kant nowhere identifies truth with provability and falsehood with refutability – or reducibility to absurdity. He also never conceived of an alternative logic. He did not reject the validity of LEM for mathematics (cf. *Jäsche Logic*, AA 9, 53). Still, through critique of the use of 'apagogic' – or *reductio ad absurdum* – proofs within metaphysics, in particular in the cosmological 'antinomies,' in the Transcendental Dialectic in CPR, it is clear that Kant did think that LEM had *limited applicability*. The trouble stems not from the fact that there are statements that we have no way of proving or refuting at present, and which therefore are neither true nor false – at least for now. A sophisticated intuitionistic take on the antinomies along these lines, in terms of empirical evidence or counterevidence – and with truth construed as warranted assertability – has been defended by Posy (1984).[3]

Rather, I shall argue that the difficulties arise from a situation in which statements are neither determinately true nor determinately false as such. This indeterminacy can best be captured through *term negation*, or so I shall suggest, where a negated predicate non-F, e.g. 'non-finite,' or 'infinite,' is on a par with F, e.g. 'finite.' There is after all symmetry between the thesis and the antithesis in the antinomies in CPR. Thus, in

---

[3]More precisely, what is ascribed to Kant is the following position. A proposition's truth consists in its eventual assertability, in the form of recognition of an effective procedure that results in evidence confirming it. In mathematics, this is combined with constructive optimism, i.e., any proposition can be constructively decided. In empirical science, there is no such constructive optimism – rather, what we are confronted with includes evidential states that progress *ad indefinitum* or *ad infinitum* – as in the 'antinomies.' This yields classical logic for mathematics and intuitionistic logic for empirical science. (See also Posy's account in the present issue of how an intuitionistic logic is part of Kant's package for empirical discourse – 'the standpoint of the Understanding.')

the judgments 'The world is finite' and the 'The world is infinite' the respective predicates are both affirmed of the subject.

It might still seem plausible, though, to ascribe a view to Kant according to which it is the outright lack of referent for 'the world' that turns both statements into falsehoods. A. W. Moore maintains this. On his construal, the reason why we are not facing contradictory propositions here is that 'the physical world' does not exist 'as a whole' (2001, 210). However, I shall argue that the subject term – 'the world' – is better seen not as vacuous but as indeterminate in what it signifies. Guided by a partial analogy Kant himself makes use of, I take this to be the reason why both the thesis and the antithesis in the First Antinomy are considered to be false. The world turns out to be neither determinately finite nor determinately infinite.[4] It has no determinate magnitude but is *incomplete* as far as this property is concerned.

Two examples from Sommers (1982, 307–327) can perhaps make it more intuitively plausible that there is room for (1) conceiving of predicates with opposite 'charge,' so that there is symmetry between two alternatives – a bit like the symmetry between z and – z; where z is a positive integer; and for (2) conceiving of a third alternative, which is neither F nor non-F. Thus, a person who lived in the past, and whose life history is therefore closed, could be either brave or non-brave. But it also seems possible that it can be indeterminate whether that person was brave or non-brave. There need be nothing in that person's character that determines this.[5] Another example is from fiction. We are not told in any of the official stories whether Sherlock Holmes met Freud or did not meet Freud. Both could have been 'facts' of fiction in their own right, but it remains indeterminate.

On the basis of this, we ought to distinguish between a propositional (or sentential) law of excluding the middle (LEM): φ or not φ, and a predicative law of excluding the middle (PLEM): Fx or (not-F)x. It is only the former that Kant accepts as a *law of thought*. PLEM is thematized implicitly in Kant, though, in so far as he puts forth a Principle of Complete Determination (PCD) for the empirical field, as we shall see. Only if PCD holds is PLEM valid. Fx or (not-F)x is of course not to be confused with Fx or not (Fx), which is just a special case of LEM. The antinomies arise, then, from attempts to force propositional negation upon lines of reasoning which

---

[4]Moore rightly acknowledges that ' ... Kant is not saying that the world in space and time does not exist in *any* sense.' (1988, 216.) But if the world exists in some sense, then surely the world-whole must exist in some sense as well.

[5]Sommers criticizes Michael Dummett on this point (1982, 311–314), who only considers the possibility that 'Jones was (not) brave' is *undecidable* – not that Jones is himself 'underdetermined to either bravery or to its opposite but this possibility is certainly coherent.' (P. 311).

do not conform to this form of negation. What is properly regarded as term negation is taken as propositional negation. While PLEM is invalid within the field of metaphysics, this is obscured by the mistaken application of LEM, which is universally valid. Here, there is a basis for limitations in LEM's applicability that lies in the subject matter of the discourse in question. Thus, we cannot use LEM to infer that x is F from x not being non-F, nor to infer that x is non-F from x not being F. In particular, we cannot prove that the world is finite nor that it is infinite in these ways. Furthermore, with the restrictions Kant put on mathematical representability, a situation where LEM does not apply in this way, and where *reductio ad absurdum* proofs therefore do not work, was not to be envisaged for mathematics. The applicability of LEM might be affected if these restrictions are relaxed, though, possibly leading to 'antinomies' within mathematics that are similar to those of metaphysics. This is arguably all that is needed to make a case for a critical view of logic within what remains a broadly Kantian conception of mathematics.

2. Formal logic as fixed by forms of thought

Let us first look briefly into Kant's notion of logic. Logic is concerned with *rules*. One reading of this is that logic is normative, in the sense that it contains rules for how one ought to think but that these rules might still be violated, i.e. there can be either logical or illogical thoughts.[6] However, this does not follow from the Kantian notion of a rule as such. Even physics is concerned with rules, according to Kant, and these are laws of material nature, not rules that can be violated. Another interpretation is therefore that the rules of logic are not normative but constitutive, i.e. they are the *necessary laws of thought*.[7] According to this interpretation, illogical thoughts are impossible. Logical 'aliens,' which think according to alternative logical laws, are also ruled out. There may still be errors because of the 'unnoticed influence of sensibility upon the understanding' (AA 9, 53). But mere thought as such is always in accordance with logical principles, though actual 'thought' need not be.

---

[6]This view is ascribed to Kant in MacFarlane (2002, 43): 'The necessary rules [of the understanding] are "necessary," not in the sense that we cannot think contrary to them, but in the sense that they are *unconditionally binding* norms for thought – norms, that is, for thought *as such*.' The view attributed to Frege, by contrast, is that the laws of logic are descriptive but not laws *of thinking*. They entail *normative rules* for thought in the sense ascribed to Kant, though. These are even called 'constitutive' norms, but they are merely taken to be constitutive of the correctness of thinking, not of thinking as such.

[7]That the laws of logic have this status is carefully argued in Tolley 2006. On a constitutive reading of the passage in the *Jäsche Logic* on the 'necessary rules' of logic (AA 9, 14), we cannot fail to think logically without failing to think. Cf. also Linnebo 2003.

Furthermore, a logic can be either *general* or *special*. A general logic concerns 'rules without regard to the difference of the objects to which it may be directed.' (CPR, A52/B76). A special logic, on the other hand, has 'rules for correctly thinking about a certain kind of objects.' The rules of general logic are necessary laws of thought as such (i.e. 'absolute'), and therefore maximally general, whereas the rules of special logic would be necessary laws for thinking about objects within some specific domain, i.e. the laws of a special science would have to be added.

Kant clearly believed that there is a special logic for each of the special sciences, i.e. for mathematics, for physics, etc. One principle that might be taken to belong to the special logic for mathematics is that of mathematical induction, as suggested by Wolff (1995, 211–9). This principle relies on the domain of natural numbers, or what is isomorphic to these. It therefore ultimately derives from the structure of the human forms of intuition – a view also ascribed to Kant in Charles Parsons's writings.[8]

The generality of general logic, by contrast, consists in the fact that its laws are valid for all thought about objects, independently of what kinds of objects these are, and even independently of the structure to be found in the human forms of intuition. Pure general logic is thus also said to be a *canon* for the *formal use* of the understanding and reason. The forms of judgment are contributed by the understanding and the forms of inference, by reason.[9] As a canon, it is a body of those principles that are necessary for the correct use of cognitive faculties. It is not an *organon*, though, i.e. a tool for acquiring knowledge. For Kant, the distinction between a canon and an organon coincides with that between a general and a special logic. Thus, whereas general logic is also called 'elementary logic,' a special logic is an 'organon of this or that science.' (A52/B76.) Hence, the special logic for mathematics – mathematical logic, we might say – is an organon for mathematics, just as the special logic for physics is an organon for physics, etc. Kant credits Lambert with having developed organons for mathematics and physics in one of the Reflexionen.[10] There is room, then, for developments within mathematical logic even in Kant.

Central to Kant's Critique of Pure Reason is *transcendental* logic. This is taken by some interpreters to be a special logic (cf., e.g. Wolff 1995, 204f.;

---

[8]'It appears that, for Kant, mathematical induction is in some way founded on the form of our intuition ... ' (Parsons 1983a, 108.)

[9]To be more precise, reason contributes the forms of syllogisms, but the understanding still contributes the forms of immediate inferences, as in conversions, e.g., from 'Some A is B' to 'Some B is A.' Cf. *Jäsche Logic* § 44 (AA 9, 115).

[10]AA 16, 48; R1629. Cf. Wolff 1995, 210.

210.) On the one hand, it is a logic that does not abstract from all content of cognition, or all relations to objects, but is exclusively concerned with pure cognitions – unlike pure general logic (A55/B80), which does not make a distinction between pure and empirical cognitions. In that sense it is specialized. On the other hand, it is still not sensitive to different kinds of objects, as are special logics in general, but 'has to do merely with the laws of the understanding and reason, but solely insofar as they are related to objects *a priori* ... ' (A57/B81-2).[11]

Kant also distinguishes between *pure* and so-called *applied* general logic. The former is what we have been speaking of so far, whereas the latter is concerned with the use of logic under empirical conditions, like shifts in attention, inclinations, etc., which can be studied in empirical psychology. Pure logic is said to have 'no empirical principles' (A54/B78). In this way, Kant at least distances himself from some crude forms of psychologism. Pure logic is not the study of actual 'thought' as such, which may be mixed rather than pure, i.e. under influences from sensibility. These are contingencies, whereas pure general logic is concerned with the necessary laws of thought. By 'logic' in Kant, I shall mean 'pure general logic,' unless otherwise stated.

Logic is *strictly topic-neutral* in the sense that there is no specific domain of objects that logic is concerned with. It abstracts from all content and all relations to objects. This does not necessarily follow from its generality, since logic could both hold of objects in general and in addition there could be a specific domain of logical objects, as in Frege. However, in Kant there is no such domain. Kant captures this property of strict topic-neutrality through his notion of *formality*.

From the fact that logic in Kant is general and formal, it follows that it deals in mere conceptual truths, which are *analytic*. Still, not all analytic principles are *logical*, and we shall see below that there are some non-logical ones that are important in mathematics.

3. From formal to transcendental logic

LEM is valid for formal logic. It is not necessarily applicable in all cases within transcendental logic, though, where '*a priori* relations' to objects

---

[11] It is also not entirely clear how to draw the distinction between a canon and an organon when it comes to transcendental logic. It is explicitly said to be a canon – not an organon. However, the *critique* is also said to be a 'propaedeutic,' as opposed to the 'system of pure reason' (A11/B25; A841/B869; A850/B878), i.e., a preparation for transcendental philosophy as itself a science. One should therefore think that there must be a special logic in the form of an organon as well, which underlies transcendental philosophy as a special science. I shall not pursue this issue further here, though.

– or relations to 'an object in general' – are also brought in. In such cases, *term negation* acquires a special significance, as we shall see, which it does not have in formal logic. This does not undermine the validity of LEM, but it brings out a source of deception in discourse that might naturally seem to be amenable to *reductio ad absurdum* proofs, and, indeed, has been taken as such in the history of thought.

Given some discourse or other that purports to be about a domain of objects, there could of course be problems with vagueness and ambiguity, which may call for sharpening or disambiguation before LEM could be applied. Kant is certainly aware of these kinds of difficulties in metaphysics, as is borne out by the 'antinomies' and 'paralogisms' of 'pure reason' in the Transcendental Dialectic in CPR. He is also aware of such difficulties in contemporary scientific discourse. For example, he criticizes certain fundamental notions in contemporary mathematics and physics which are not sufficiently clarified. Kant no doubt holds that these flaws can be eliminated through his prescribed methods for the exact sciences.

A more fundamental problem remains. Assume that there is some specific non-empty domain of objects, and let **a** be an object within this domain. It does not follow analytically from its being either the case that F**a** or its not being the case that F**a** that either F can be predicated of **a** or non-F can be predicated of **a**. That is, the Principle of Complete Determination, PCD – which is required for PLEM to be valid – cannot be derived from LEM.[12]

Now, one reason to claim that x is neither F nor non-F could be that these are not predicates that 'span' x, to borrow a term from Sommers. Thus, a number is neither red nor non-red, because neither can be sensibly predicated of numbers. We shall see below that Kant does arguably have considerations to this effect, which puts limits on what can sensibly be predicated of objects on the basis of what kind of objects those are. However, it could also be that even though a pair of predicates F/non-F can sensibly be predicated of x, x is neither determinately F nor determinately non-F. There is the possibility, namely, that x is *incomplete* with regard to the property designated by F, i.e. it might still be indeterminate whether x is F or whether it is non-F.

---

[12] Note that 'predicate' here can be either a predicate concept or a linguistic expression for such a concept. According to Kant, there can never be a concept without a word for it. See Michael Wolff 1995, 24, and, e.g., *Anthropology*, AA 7, 167 & 192; *Vienna Logic*, AA 24, 934. This is of particular relevance when we turn to construction in mathematics, which cannot be 'an essentially languageless activity of the mind,' as Brouwer would have it.

The deeper problem here is that of incompleteness. In such cases, we cannot use LEM and *reductio ad absurdum* proofs to establish whether x is F or whether it is non-F. After all, not (not-F)x does not entail Fx and not Fx does not entail (not-F)x. The opposition between x being F and x being non-F is of the propositional form φ or ψ – it is not of the propositional form φ or not φ.

Kant's distinction between the logical forms of *negative* and *infinite* judgments is relevant here. To use his own example from CPR, there is an important difference between the judgment 'The soul is not mortal,' and the judgment that 'The soul is non-mortal.' To claim that the soul is not mortal is merely to exclude it from the sphere of everything mortal. On the other hand, to claim that the soul is non-mortal is to make a claim that is positive according to its form, namely, that the soul belongs to the sphere of everything that is non-mortal. Whereas the infinite judgment has existential commitment, the negative judgment does not. The distinction is therefore important in transcendental logic (cf. CPR, A72/B97) but not in formal logic, which abstracts from all relations to objects.

There are in fact two ways of construing infinite judgments here, namely, *with or without range restriction*. With range restriction, the sphere of the possible has two subspheres, that of mortal beings and that of non-mortal beings. However, the sphere of the possible is not thereby exhausted, as there are other subspheres of objects which are not spanned by the predicates 'mortal'/'non-mortal,' like, e.g. geometrical figures. Without range restriction, on the other hand, the sphere of the possible is divided into only two subspheres by the predicates 'mortal'/'non-mortal,' those beings which are mortal and the rest.

In either case, with or without range restriction, it is clear that if the Principle of Complete (or Thoroughgoing ['durchgängig']) Determination (PCD) holds for judgments about an object, it guarantees that that object definitely belongs within one sphere or another. Thus, let x be any object and let F be a predicate that can sensibly be predicated of x. Without adding PCD to the principles of pure general logic, there is the possibility that an x which is excluded from a sphere is not thereby included in another sphere, i.e. that even if it is not F, it is also not non-F, as it could be neither determinately F nor determinately non-F.

Kant brings up this principle in the Transcendental Ideal in CPR, and he makes it clear that it 'deals with the content and not merely the logical form.' (A572/B600.) It belongs, therefore, to transcendental logic, not to general logic, and Kant says explicitly, in the same passage, that it 'does not rest merely on the principle of contradiction ... '

As we shall see below, we may have to accept incompleteness within certain areas. If incompleteness is to be ruled out within an area, we have to presuppose that there is a ground for complete determinacy within that area, so to speak, that makes it the case that either F or non-F for all F's that can sensibly be predicated of x. In the case of objects of experience, coherence according to causal laws would have to play a part in this.[13] I shall not discuss the empirical case here. We shall ask, rather, what could possibly ensure complete determinacy in the mathematical case. I shall suggest that according to Kant we cannot have constructive access to anything which is not pre-determined, so to speak, by the structure of our human forms of intuition. *It is the structure that is contained within space and time as forms of intuition that is the ground for complete determinacy, and thus what makes a mathematical statement determinately true or determinately false.*[14] If we allow mathematics to go beyond what is pre-determined by this structure, the validity of PCD will no longer be guaranteed. This brings us to the issue of constructions in mathematics.

4. From logical principles to principles of construction in mathematics

Logic as fixed by mere forms of thought is general and formal. Transcendental logic brings in content and, thus, possible range restrictions and a completeness presupposition. It remains general in the sense that it does not discriminate between objects as such. Mathematics, on the other hand, is neither general nor formal.

As for generality, space and time as human forms of intuition are the scenes for mathematical construction in Kant. Our mathematics, therefore, need not obtain for objects which are not given or givable in our forms of intuition. There may be room, though, for an abstract arithmetic that holds even for things in themselves, in so far as there could be an abstract notion of number, in accordance with the pure categories of quantity. This has been discussed by Parsons (2012, 57–68), but I shall not pursue the issue further here.

More important than generality in our context is the fact that mathematics is not strictly formal, in the way formal logic is. It does not

---

[13]Cf. the discussion in Stang 2012. I disagree with some of his further considerations, though, which lead him to conclude that empirical objects' conformity to PCD – with regard to their empirical properties – is not an a priori matter for Kant.

[14]This position has been worked out in detail in the form of a comprehensive Kantian theory of mathematical construction in Sandmel 2001.

concern mere forms of judgments and inferences that are fixed by the nature of our understanding and reason. It does not, therefore, abstract from all relations to objects but rather comes with its own domains of objects. The objecthood in question is not a thick one, though. Rather, '[a]n *object* (...) is that in the concept of which the manifold of a given intuition is *united*.' (B137.) And construction of mathematical concepts is a way of uniting – or synthesizing – the manifold that is given in the medium of pure intuition. Moreover, the synthesis in question is arbitrary, and here lies the potential in mathematics for developing new concepts with content. A domain of objects that is explicitly acknowledged by Kant is of course that of geometrical figures, but arguably there are also other domains of objects, like that of finite sets and numbers – or abstract quantities, since manifolds of intuition are united under these concepts, like manifolds of strokes in an arithmetic with stroke strings for the numbers.

Given Kantian 'constructivism,' one might think that all evidence in mathematics has to be 'constructive.' However, Kant in fact accepts both *ostensive* and *apagogic* – or *reductio ad absurdum* – proofs in mathematics. There is even a special kind of self-evidence that comes with a contradiction, i.e. 'clarity of representation' to a high degree (A790/B818; cf. A789-92/B817-20). On the other hand, we have a deeper grasp through 'constructive' evidence, or some form of 'ostensive proof.' Why is this so?

Mathematical statements are not true or false about objects that exist independently of human conceptions. There is no Platonism in this sense in Kant. However, mathematical statements are also not about constructions as such. For Brouwer, it is neither true nor false that there are seven consecutive 7's in the decimal expansion of π as long as we have no guarantee that this pattern is there – by calculation that has proceeded sufficiently far – nor any guarantee, by way of a theorem, that it will not be found.

For Kant, on the other hand, a potential sequence is not a mathematical object in its own right – 'proceeding' or 'growing' over time. Still, if the rule for a sequence is given, arbitrary initial segments are in principle constructible. But the 'truth-maker' for mathematical statements – or that by virtue of which they are true – are not constructions but structure which is in some sense 'contained in' our human forms of intuition prior to all constructions. In this pre-given structure, π is already completely pre-determined, and we can even have constructive access to it as such, namely, in the form of the ratio between a pair of lengths – the circumference of a circle and its diameter.[15] Any such property of a sequence of digits in the decimal expansion for π is therefore also already completely pre-determined.[16] Thus, in these cases PCD holds, PLEM is valid, and LEM is applicable. Even if we do not yet know whether π has the property designated by F or the property designated by non-F, we know that it is either F or non-F, as validated by PLEM. Whether such pre-determination can be generalized to properties that involve nested quantifiers or impredicativity is of course another matter. Kant just presents a general view of mathematics as such, according to which mathematical properties 'flow' from arbitrary syntheses of intuitive manifolds. These syntheses are determined by conceptual rules introduced by the mathematician.

It is not sheer rational optimism, then, but the view that mathematical structure is in a sense already contained in the human forms of intuition that underlies Kant's belief that mathematics is decidable – i.e. that all meaningful mathematical statements are either provable or refutable. This belief was to be echoed later, by Hilbert's 'no ignorabimus.' In Kant, this is an instance of a more general belief. In so far as reason is under strict structural constraints, all the questions it poses are answerable one way or the other, and the answer is even pre-determined by the structural constraints in question. Were we to go beyond these constraints, however, there would no longer be any guarantee of solvability.

> It is not as extraordinary as it initially seems that a science can demand and expect clear and certain solutions to all the questions belonging within it (*quaestiones domesticae*), even if up to this time they still have not been found. Besides transcendental philosophy, there are two pure sciences of reason, one with merely speculative, the other with practical content: *pure mathematics* and *pure morals*. Has it ever been proposed that because of our necessary ignorance of conditions it is uncertain exactly what relation, in rational or irrational numbers, the diameter of a circle bears to its circumference? Since it cannot be given congruently to the former, but has not yet been found through the latter, it has been judged that at least the impossibility of such a solution can be known with certainty, and Lambert gave a proof of this. (CPR, A480/B508.)

So pure rational sciences are decidable. They are 'sciences whose nature entails that every question occurring in them must absolutely be answerable from what one knows, because *the answer must arise from the same source as the question ...*' (A476/B504; my italics). Besides mathematics,

---

[15]Cf. Kant's letter to Rehberg, dated some time before September 25, 1790 (AA 11, 207-210), where $\sqrt{2}$ is considered as determined by the ratio between the diagonal and the side of a square.

[16]Cf. Sandmel 2001, 544–5, on how the decimal expansion of pi is 'synthetically determined' by way of 'pure manifold structures,' and, hence, the sequence '0123456789' contained therein was *discovered* somewhere after its 17 billionth digit by Kanada in 1997 – it was not invented.

transcendental philosophy and ethics also have this status. I shall not go into ethics, but we shall see below that *reductio ad absurdum* proofs can indeed be used within transcendental philosophy, which is methodologically clarified in a way that the special metaphysics of the tradition is not.

In the quoted passage, Kant's brings up a particular mathematical example. When Lambert presented a proof – in the 1760s – which showed that π is an irrational number, Kant took this to confirm his belief that mathematics is decidable. The proof makes use of a continued fraction and thus of *iteration*, which is central in Kant's 'constructivism.' Kant's further remark possibly hints at the question of whether π is constructible with compass and ruler – i.e. whether it is an algebraic or a transcendental number.[17] Again, Kant has a firm belief that an answer to this question will be found. It did not happen in his lifetime, though. The first one to publish a proof that π is transcendental was Lindemann in 1882.

Kant's interest in such issues of decidability within mathematics is noteworthy. The status of mathematics as a pure rational science – more than its potential for conceptual innovations – was after all crucial for the Kantian project, since mathematics was to serve as a model for metaphysics itself as set on 'the secure course of a science,' (Bvii; cf. Bxv-xvi) i.e. as a pure rational science, in the form of transcendental philosophy.

According to Kant's position, then, mathematics is transparent to reason in the sense that no mathematical structure is in principle hidden or inaccessible to us. Furthermore, all mathematical statements are either determinately true or determinately false. On the basis of this belief in the completeness of mathematics, Kant is convinced that all mathematical problems are solvable – we will eventually find a proof or a disproof.

Still, Kant does think that there are problems posed by pure reason *within metaphysics* that are not decidable one way or the other. To throw light upon how constraints on constructibility in mathematics is supposed to ensure decidability, we shall therefore look into such undecidable problems – which are to be found in the Transcendental Dialectic in CPR. We then have to ask ourselves whether there could be analogues to problems like these within mathematical discourse after all, if we relax the more specific constraints that Kant has put on mathematical construction.

---

[17]This is suggested in Büchel 1987; 354–5. Cf. also Martin 1985, 57–8.

5. The false presupposition of the 'mathematical antinomies'

For transcendental philosophy itself to become a rational science with decidability, metaphysics had to be subjected to a critique of pure reason. Traditional metaphysics was divided into *metaphysica generalis* and *metaphysica specialis*. The former was general ontology, which is transformed into an Analytic of Principles in Kant, whereas the latter is dealt with in the Dialectic. The CPR presents a propaedeutic to transcendental philosophy as a rational science and thereby also what Kant calls a 'metaphysics of metaphysics' in a letter to Marcus Herz, dated ca. May 11, 1781 (AA 10, 268-270). At this meta-level, LEM and *reductio ad absurdum* proofs are accepted. The Antinomies of Pure Reason can be read as an indirect proof for transcendental idealism over transcendental realism – there is no third alternative at this level of clarification. With the position of transcendental idealism, each of the antinomies is in a way dissolved, or at least its deceptive character is removed. At the object-level of special metaphysics, on the other hand, which is concerned with very special 'objects' – the world, the soul and God, *reductio ad absurdum* proofs will not work. Let us look into why this is so.

With a view to what might be relevant for mathematics in particular, we shall concentrate on the first and second pairs of thesis and antithesis, i.e. on those labelled 'mathematical' antinomies. They deal with the world in its aspects of extension and composition. Kant in one place says that the arguments in the antinomies are *valid according to their mere form*. I think what is meant is that they do contain the valid form of a reductio in them – a valid form that belongs to pure general logic, i.e. a particular use is made of this form. Still, the transition from not (not-F)a to Fa and from not Fa to (not-F)a are not legitimate. They are in fact not inferences from the absurdity of not φ to φ, as legitimized by LEM – the propositional law of excluding the middle. Rather, they are ultimately invalid because they rely on the PCD of transcendental logic. Transcendental logic is after all not merely formal, unlike general logic. It turns out that PCD does not hold here, and PLEM – the predicative law of excluding the middle – is invalid.

In the First Antinomy, the thesis is that the world is finite in extension (in space and time), whereas the antithesis is that the world is infinite in extension (in space and time). In the Second Antinomy, the thesis is that matter has ultimate constituents, whereas the antithesis is that matter is infinitely divisible, and thus without ultimate constituents. In the solution to the antinomies, Kant argues that both thesis and antithesis are false. How is that possible?

Kant does not take himself to reject LEM as such. He makes a distinction between 'analytical' and 'dialectical' opposites (cf. A504/B532). In the former case, we have real contradictories. In the latter case, we do not:

(1) The world is finite in extension
(2) The world is not finite in extension
(3) The world is non-finite, or infinite, in extension

1 and 2 are analytical opposites, whereas 1 and 3 are dialectical opposites. It is possible that both 1 and 3 are false, but not both 1 and 2.

One might maintain that the arguments are therefore *merely apparent reductio ad absurdum* proofs. This does not show that LEM is not a necessary law of thought, but it does show at least that proofs based on this principle cannot be given for the claims that are made here, at the object-level of special metaphysics, so to speak. More generally, what is suggested is that LEM cannot be put to any good use in dialectical thought, as opposed to analytical thought. It is not invalid here, but it is inapplicable in solving the problem of whether the world is finite or infinite. This is so because neither F nor non-F can be established of x through *reductio ad absurdum* proofs.

To explain what goes wrong here, Kant makes use of an example (cf. A503/B531):

> All bodies have either a good smell or a smell that is not good, or, equivalently, all bodies are either good-smelling or non-good-smelling.

As Kant points out, there may be bodies which do not have a smell at all. So, from the fact that a body is not non-good-smelling, we cannot infer that it is good-smelling; and from the fact that a body is not good-smelling, we cannot infer that it is non-good-smelling.

This seems to suggest a *range restriction*, then, in the case of the pair of predicates 'good-smelling' and 'non-good-smelling.' Kant argues that similarly, from the fact that x is not infinite, or not non-finite, we cannot infer that it is finite. And from the fact that x is not finite, we cannot infer that it is infinite, or non-finite. It may neither be the case that x is finite nor that it is infinite, i.e. x may not be determinate in its magnitude at all. According to Kant, that is the case with the world.

The parallel is misleading in at least one important respect, though. While 'good-smelling'/'non-good-smelling' is a pair of predicates that span only those objects that have a smell, 'finite'/'infinite' is arguably a pair of predicates that span all objects whatsoever. So, if neither x is finite nor x is infinite is true, then that is not because of a range restriction on the pair of predicates 'finite'/'infinite.' There must be another reason, namely, the *false presupposition of completeness* – or determinacy in properties – which underlies the mathematical antinomies.

6. The 'constructive' solution to the mathematical antinomies

The reasoning in the First Antinomy proceeds according to mere conceivability and successive synthesis in pure imagination. It does not bring in empirical evidence as such. After all, the world is not an object to be met with in experience on a par with other objects. Instead, the reasoning is in effect subjected to the constructibility conditions of the understanding. Let us look at the arguments against the finitude and the infinity of the world in space, respectively.

A finite world would have definite boundaries in space, which is a determination relative to pure space itself. It would also be in definite states of motion, which is again a determination relative to pure space itself. Both are impossible, for pure space is no *thing* to which anything can be *related*. In construction of a finite geometrical space, there is no absolute boundary provided by pure space, and in construction of a state of motion there is no absolute frame of reference provided by pure space. A determinately finite world in an otherwise empty space is therefore not conceivable.

An infinite world as a collective totality in space is not structurally similar to anything that is geometrically constructible or determinately measurable. Constructed line segments or measured distances are always finite, even if there is no upper bound on their length. Furthermore, construction *ad infinitum* in geometry – say by iterated extension of a line segment – proceeds according to a rule that guarantees the next step in infinite space, but there is no such rule when it comes to the measurement of the world's extension. While we cannot run out of space, we may run out of matter, or we may not. A determinately infinite world is therefore also not conceivable.

Whatever the status and merits of these 'constructivist' arguments as far as the physical world is concerned, what is of particular interest here is the rejection of an actually infinite collective totality. It belongs to Kant's considered view of infinity, and it throws light upon which specific principles of construction he accepts in mathematics. It is *conceptually impossible* that there be such infinity in the synthesis of a manifold. Kant presents both a 'transcendental' and a 'mathematical' concept of infinity, which are taken to be co-extensional from the standpoint of a human understanding:

> The true (transcendental) concept of infinity is that the successive synthesis of unity in the traversal [Durchmessung] of a quantum can never be completed. (CPR, A432/B460.)

> This [quantum] thereby contains a multitude [Menge] (of given units) that is greater than any number, and that is the mathematical concept of the infinite. (CPR, A432/B460n.)

What is infinite in mathematical construction can only be generated successively, and this is *potential*, not actual, infinity. *Still, the restriction to potential over actual infinity in mathematical construction does not lead Kant to seek an alternative logic* – though a route to mathematical intuitionism lies precisely in considerations of the status of the infinite. As we saw in Section 3, Kant takes the solution to a mathematical problem to be predetermined by the structure of our forms of intuition, and, thus, to hold independently of what has so far been traversed in the form of actual constructions and proofs.

On the other hand, it might seem as if even these objections to actual infinity can be circumvented if we accept that there can be numbers for the cardinal size of actually infinite multitudes, and that each of these multitudes need not be constructed ostensively. Symbolic construction might be all that is required. While a number for Kant is always a natural number, the two concepts of infinity would no longer be co-extensional if numbers for actually infinite multitudes were also acknowledged. Rather, there would be numbers for multitudes which are no longer mathematically infinite – according to Kant's characterization thereof.

It is hard to maintain this, though, without considering these multitudes as pseudo-traversed, or as quasi-completed in their formation, and thus not as infinite in an absolute sense, either, but rather as 'trans-finite' in cardinal size, in a sense akin to Cantor's. Indeed, if quasi-completion in stages is accepted as 'successive synthesis' of sorts – so that the notion of such synthesis is very much liberalized – this could transform the transcendental and mathematical concepts of infinity into co-extensional concepts of the absolutely infinite – which is unlimited even by quasi-completion and transfinite numbers, respectively. This would call for a radical reconceptualization, then, of infinity. Let us look at where this might lead us when it comes to a Kantian conception of mathematics.

7. Are there 'antinomies' within mathematics itself?

It is certainly possible to see Kant's critique of an all-inclusive totality in the Antinomies as paralleled by critique of attempts to complete an essentially incompletable totality from the perspective of iterative set theory. Ernst Zermelo spoke of a 'formal analogy' in his edition of Cantor's collected works (Cantor 1932, 377). Moore suggests that a 'dilemma can be presented, in Kantian style' (Moore 1988, 217). Let the Universe be the set of all sets. Let the Thesis be that the Universe belongs to itself and the Antithesis that it does not. Indirect proofs for both Thesis and Antithesis, respectively, can then be given. If the Universe does not belong to itself, there is at least one set that it does not contain, namely, the Universe. If the Universe does belong to itself, it is not a set, which can only contain sets of lower rank. As in the discussion of the cosmological antinomies above, we may let F be the property in question. It is then not the case that x either has the property F or the property non-F, when x is the set of all sets. There is therefore effectively a basis for a critique of the application of LEM here, in the way it might be put to use within set theory. Still, there can be something like a regulative idea of the Universe – just as there is a regulative idea of the physical universe in Kant's cosmological antinomies – so that one goes on 'exploring axioms that postulate bigger and bigger sets. We are to proceed *as if* there were a set of all sets.' (Moore 1988, 217.)

Unlike Moore, I do not construe this parallel as resting on an outright lack of referent for 'the physical world as a whole,' though, as noted in the Introduction above. If there were to be such a lack of referent, it would have to be by necessity, not contingently. Only then could we have a priori insight into the matter. A contradictory concept lacks a referent by necessity, and we know this a priori, but the idea of the world-whole – a 'concept' of reason – is not contradictory, which is precisely why it can have a regulative role. While it is merely *as if* there is a world-whole with a determinate magnitude, the idea of a world-whole is not contradictory as such. I construe the resemblance, therefore, as incompleteness, or essential incompletability, in what 'the world-whole' signifies.[18] It is term negation together with the suspension of PCD

---

[18] Although Moore acknowledges that the world exists in some sense for Kant, his construal actually fits better with what I take to be Kant's simplification of his own position, as presented in *Prolegomena* (§52b, AA 4, 341). There, the pair of thesis and antithesis is likened to the following pair of propositions: (1) A square circle is round; (2) A square circle is not round. It is noteworthy that in this case, Kant does think that we have an analytical opposition between 1 and 2, i.e., they are real contradictories: '... since between these two [propositions] no third proposition can be thought, through this concept *nothing at*

that makes this option available to us within transcendental logic, as we have seen.

To be sure, mathematics is from the very outset subject to strong restrictions in Kant, unlike rational cosmology – which can easily become too speculative. The restrictions put on mathematics are boldly transgressed in totalizing quasi-constructions of reason. The constraints come both in the form of synthetic principles of construction and analytic principles like 'the whole is greater than its [proper] part,' (cf. B16-7) which is known as *Euclid's principle* – one of the five 'common notions' in Euclid. It does not belong to mere logic as such, but it is analytic. As self-evident and indemonstrable, it can be taken as an 'axiom' in an Aristotelian sense but not in a Kantian sense. For Kant, axioms are always synthetic.

In so far as analytic non-logical principles are taken to be concerned with equivalence and with the part/whole relation generally, they can be considered as valid for all the mathematical sciences, and not just for geometry. They show the extent to which conceivability by the understanding, besides intuitability, plays an important part in restricting the sphere of mathematical possibilities in Kant. However, given the totalizing quasi-constructions of reason, it is still possible that there could be a clash between two such principles.

For example, three principles of *equinumerosity* can be derived from Euclidean common notions – those of equality in cases of equals added to or subtracted from equals, respectively – together with an additional principle of 'Just as many as none.' (cf. Sutherland 2017, 185–6). The trio of principles can plausibly be considered to follow analytically from the concept of equinumerosity in question. The trio also entails the principle of one-to-one correspondence for finite collections. Arguably, the latter, which does the same work as the former, is analytic as well. Furthermore, while 'pairing off' acts as a determinate 'measure' by which two finite sets are of the same cardinal size, it is very 'natural' for reason to extend this from finite to infinite cases, which also leads to an extended principle of conceivability that conflicts with Euclid's principle.

Historically, we may use Galilei's paradox to illustrate this. Kant probably knew of the part/whole paradoxes that were discussed in the Middle Ages, and also of the one that was presented in Galilei's work *Dialogues Concerning Two New Sciences* of 1638 (cf. Galilei 1954, 31-33). Be that as it may, the conflict between two principles that should be on a par as principles of

---

all is thought.' (AA 4, 341.) He goes on to claim that what underlies the mathematical antinomies is precisely 'a contradictory concept of this type ... ' A square circle is of course not constructible – its concept is outright contradictory and there is no referent for its concept.

conceivability can be brought out as follows. On the one hand, the set of squares is a proper subset of the set of natural numbers. By Euclid's principle, the set of natural numbers is therefore greater than the set of squares. On the other hand, by the 'measure' of one-to-one correspondence, there are as many squares as there are natural numbers.

This is an 'antinomy' of reason within mathematics itself. The fact that the assumption of an actually infinite set leads to this antinomy does not itself curb the totalizing quasi-constructions of mere reason, as it were. Kant would also be familiar with positions according to which there can be infinities of different size in mathematics. After all, such a view was defended by Schultz (1788, § 34) – Kant's most trusted expositor, albeit without a Cantorian notion of infinite powers.[19] However, if we give up on the finitary constructibility condition, it is possible that we end up with illusory completeness in set theory also, just as in the cosmological antinomies in CPR. Subjective conditions – reason's search for creative progress as well as for closure or completeness – can be conflated with objective properties. This is so if we have sufficiently liberal principles of *symbolic* – as opposed to *ostensive* – construction and accept quasi-completion of infinite collective totalities through these. While the abstract Universe remains a good candidate for illusory ascriptions of completeness, an 'actually' infinite set as such is arguably not, given the Cantorian line of developments after Kant's time. It has become widely accepted that there can be both finite and transfinite sets with definite size. They are completable or quasi-completable in their formation and therefore also complete in the sense of being determinate in their magnitude, i.e. with regard to a property that belongs to the wholes as thus formed.

8. Conclusion: Quasi-constructions of reason and the status of LEM

To see that there are resources in Kant's philosophy for rejecting his own finitary constructibility condition in favour of a trans-finitary one, we should look at the reason why there are antinomies *within metaphysics*. It is natural for us to take the physical universe as a complete whole, in accordance with the First Antinomy – i.e. as a 'large thing' in its own right – but it is not. The world as an all-inclusive totality is not wholly

---

[19]There are different orders of infinity, though, and it is noteworthy that his theory is not simply based on Euclid's principle. Rather, Schultz offers a critique of this principle. Whether the whole is indeed greater than the proper part when both are *infinite* magnitudes has to be decided by other means (cf. Schultz 1788, 254).

traversable or pseudo-traversable – the way we have construed sets as completable or quasi-completable. Nor need it be complete with regard to properties of the whole. In particular, it is incomplete in that respect in not being determinate in its magnitude, i.e. it has neither the property of being finite nor that of being infinite. What, then, is the root of its projected completeness? According to Kant, the cosmological antinomy is due to a *subreption*, i.e. a merely subjective condition is being conflated with something objective. The situation in mathematics is different.

> Apagogic proof, however, can be allowed only in those sciences where it is impossible *to substitute* that which is subjective in our representations for that which is objective, namely the cognition of what is in the object. (...) In mathematics this subreption is impossible; hence apagogic proof has its proper place there. (CPR, A791-2/B819-20.)

Kant's official view is thus that there can be no subreptions in mathematics. It easily follows that PCD is valid for mathematics. We do not just project completeness in properties where none is to be found.

Moreover, in transcendental philosophy antinomies serve as an indirect proof of transcendental idealism. There is no equivalent to this in mathematics. We cannot be transcendental idealists about mathematics because it deals with structures themselves, and not with what is passively received and structured in one way or other. It is the latter that gives rise to the distinction between phenomena and noumena,[20] and to the possible conflation of the one with the other – as when one seeks ultimate constituents in infinitely divisible matter, in accordance with the Second Antinomy.

Kant did of course not foresee the developments within set theory. But he seems to have had an influence on Zermelo, who acknowledged two strong tendencies in the human mind that underlie Kantian antinomies. This is beautifully expressed in a pioneering paper from 1930, 'On boundary numbers and domains of sets: New investigations in the foundations of set theory.'

> The two diametrically opposed [polar entgegengesetzten] tendencies of the thinking mind, the ideas of creative *progress* [schöpferischen *Fortschrittes*] and collective *completion* [zusammenfassenden *Abschlusses*], which form also the basis of the Kantian "antinomies," find their symbolic representation as well as their symbolic reconciliation in the transfinite number series, which rests upon the notion of well-ordering and which, though lacking in true completion on account of its boundless progressing [Schrankenlosen Fortschreiten], possesses

---

[20] Posy makes an apt distinction between evidence with external origins, in the empirical sciences, and evidence with internal origins, in the rational sciences (cf. Posy 1984, 128).

relative way stations [Haltpunkte], namely those "boundary numbers" [Grenzzahlen], which separate the higher from the lower model types. (Zermelo 2010, 431; transl. slightly modified, cf. the original on p. 430.)[21]

Zermelo rejects Kant's theory of mathematics as based on pure intuition ['reine Anschauung'], but Kant's theory of antinomies 'expresses a deeper insight, an insight into the "dialectical" nature of human thinking.'[22]

With the acceptance of principles of construction based on this natural dialectic, there is no purported guarantee, so to speak, in our human forms of intuition, for completeness anymore. This may be so in so far as *reason's own constructions*, progressing through higher levels of infinity, as in set theory, take the place of constructions restricted by the understanding. It calls for idealizations in mathematical construction that go well beyond anything Kant envisaged. Still, as we know, such idealizations became part of mainstream mathematics, and Kant had a conservative, anti-revisionary attitude to the mathematics of his time. It is therefore very much in the Kantian spirit to discuss the status or role of LEM in statements that one encounters in present-day mathematical discourse, and not just Kant's own examples of mathematical statements.

Mathematics as developed in agreement with its own *organon* does not adhere to Kant's finitary constructibility condition, nor does it respect all the analytic non-logical principles of mathematics that are suggested by his condition of conceivability through the understanding, like Euclid's principle. This raises the following possibility: Even given appropriate range restrictions, there may be cases where LEM is inapplicable because the completeness presupposition does not hold and PCD fails. A *broadly Kantian* conception of mathematics, according to which its characteristic feature is manipulations of signs given in intuition, does not rule this out. Indeed, mathematical antinomies might be expected to emerge once we transcend the structural constraints that come from pure space and time intuition. Within such a liberalized 'intuitionism' and 'constructivism,' there is no mathematical Platonism that might serve as a substitute for pre-determination through the forms of intuition, so as to secure completeness and consistency. Hence, reflections on constructibility conditions and introduction of other principles of construction

---

[21]This passage is also quoted in Kanamori 2004, 529.
[22]This is quoted in Ebbinghaus and Peckhaus 2015, 203n, from Cantor's Collected Papers (1932, 377). Zermelo saw a parallel between the set-theoretic antinomies and Kant's antinomies of pure reason. Zermelo used to point out such affinities in his courses on set theory, according to Gottfried Martin (1955, 54), who attended his lectures.

in mathematics than those Kant himself laid down may leave room for a Kantian critique of logical principles, like LEM. In particular, just as reflections on metaphysics provide a basis for such a critique in Kant, we have seen that there may be a basis in set theory for a distinctly Kantian critique of classical logic.[23]

Disclosure statement

No potential conflict of interest was reported by the author.

References

Büchel, Gregor. 1987. *Geometrie und Philosophie*. Berlin: Walter de Gruyter.
Cantor, Georg. 1932. *Gesammelte Abhandlungen*. Edited by Ernst Zermelo. Berlin: Springer.
Ebbinghaus, Heinz Dieter, and Volker Peckhaus. 2015. *Ernst Zermelo. An Approach to His Life and Work*. 2nd edition. Berlin: Springer.
Galilei, Galileo. 1954. *Dialogues Concerning Two New Sciences*. Translated by Henry Crew and Alfonso de Salvio. New York: Dover.
Kanamori, Akihiro. 2004. "Zermelo and Set Theory." *The Bulletin of Symbolic Logic* 10: 487–553.
Kant, Immanuel. 1900. *Gesammelte Schriften*. Königliche Preussische (later Deutsche) Akademie der Wissenschaften (ed.). Berlin: Georg Reimer, later Walter de Gruyter. Cited as *AA*.
Kant, Immanuel. 1998. *Critique of Pure Reason*. Edited by Paul Guyer and Allen Wood. Cambridge: Cambridge University Press. Cited as CPR.
Linnebo, Øystein. 2003. "Frege's Conception of Logic: From Kant to Grundgesetze." *Manuscrito* 26: 235–252.
MacFarlane, John. 2002. "Frege, Kant, and the Logic in Logicism." *The Philosophical Review* 111 (1): 25–65.
Maddy, Penelope. 2007. *Second Philosophy. A Naturalistic Method*. Oxford: Oxford University Press.
Martin, Gottfried. 1955. *Kant's Metaphysics and Theory of Science*. Translated by P. G. Lucas. Manchester: Manchester University Press.
Martin, Gottfried. 1985. *Arithmetic and Combinatorics: Kant and His Contemporaries*. Translated by J. Wubnig. Carbondale and Edwarsville: Southern Illinois University Press.
Moore, Adrian W. 1988. "Aspects of the Infinite in Kant." *Mind* 97: 205–223.
Moore, Adrian W. 2001. *The Infinite*. 2nd edition. London: Routledge.
Parsons, Charles. 1983a. "Infinity and Kant's Conception of the "Possibility of Experience"." In *Mathematics in Philosophy*, 95–109. Ithaca: Cornell University Press.
Parsons, Charles. 1983b. "Kant's Philosophy of Arithmetic." In *Mathematics in Philosophy*, 110–149. Ithaca: Cornell University Press.

---

[23] I am grateful to Mirja Hartimo and Øystein Linnebo for comments on earlier drafts of this paper.

Parsons, Charles. 2012. "Arithmetic and the Categories." In *From Kant to Husserl*, 42–68. Cambridge, Mass.: Harvard University Press.

Parsons, Charles. 2015. "Infinity and a Critical View of Logic." *Inquiry* 58 (1): 1–19.

Posy, Carl. 1984. "Kant's Mathematical Realism." *The Monist* 67 (1): 115–134.

Sandmel, Thor. 2001. *Matematisk erkjennelse. Grunnriss av en transcendentalfilosofisk fundert matematikkfilosofi*. PhD-thesis. Oslo: Unipub.

Schultz, Johann. 1788. *Versuch einer genauen Theorie des Unendlichen*. Königsberg and Leipzig: G. L. Hartung.

Sher, Gila. 2016. *Epistemic Friction. An Essay on Knowledge, Truth, and Logic*. Oxford: Oxford University Press.

Sommers, Fred. 1982. *The Logic of Natural Language*. Oxford: Clarendon Press.

Stang, Nicholas F. 2012. "Kant on Complete Determination and Infinite Judgement." *British Journal for the History of Philosophy* 20/6: 1117–1139.

Sutherland, Daniel. 2017. "Kant's Conception of Number." *Philosophical Review* 126 (2): 147–190.

Tolley, Clinton. 2006. "Kant on the Nature of Logical Laws." *Philosophical Topics* 34: 371–407.

Wolff, Michael. 1995. *Die Vollständigkeit der kantischen Urteilstafel*. Frankfurt am Main: Vittorio Klostermann.

Zermelo, Ernst. 2010. *Collected Works*. Vol. 1. Edited by Ebbinghaus, Heinz-Dieter, Craig G. Fraser and Akihiro Kanamori. Berlin: Springer.

The infinite, the indefinite and the critical turn: Kant via Kripke models

Carl Posy

ABSTRACT
This paper aims to show that intuitionistic Kripke models are a powerful tool for interpreting Kant's 'Critical Philosophy'. Part I reviews some old work of mine that applies these models to provide a reading of Kant's second antinomy about the divisibility of matter and to answer several attacks on Kant's antinomies. But it also points out three shortcomings of that original application. First, the reading fails to account for Kant's second antinomy claim that matter is divisible 'ad infinitum' and his first antinomy claim that empirical space extends only 'ad indefinitum'. Secondly, in conformity with the 'assertability' heuristics for Kripke models, it attributes an assertability semantics to Kant. In fact, it attributes a pair of assertability semantics to Kant, but neither of these accords with modern assertabilism. And finally, one must wonder the propriety of using contemporary assertability and contemporary formal logic to save an eighteenth-century text. Part II goes historical to solve the problems about assertability. It outlines the Descartes–Leibniz dialectic about the infinite/indefinite distinction and demonstrates how each position reflects larger philosophical positions. It then demonstrates that Kant's two versions of assertability and his version of the infinite/indefinite dichotomy emerge naturally from his attempts (pre-critical and then critical) to resolve a tension in Leibniz's philosophy. In light of this demonstration, Part III adapts the Kripke model interpretation so that it does reflect that dichotomy. It then refines the Kripke semantics in order to model some of Kant's signature critical doctrines, and it shows how Kant's critical version of the infinite/indefinite distinction differs from all those of his predecessors and from his own pre-critical version. It also addresses the question of using contemporary logical machinery to interpret Kant. Part IV turns the tables and uses what we have learned about Kant and modern semantics in order to address a pair of issues in the contemporary critical foundations of logic.

I thank the editors for inviting me to contribute to this issue on critical views of logic. Kant invented the *critical* philosophy. He fashioned its doctrines (Understanding versus Reason, synthetic *a priority*, transcendental idealism, empirical realism) into a philosophy designed to replace the dogmatic realism of his predecessors and of his own 'pre-critical' thought. His antinomies used *logic* to reduce that realism *ad absurdum*. The realist of the first antinomy says that the physical world is either finite or infinite in extent and yet must deny each disjunct; and similarly the second antinomy realist affirms that there are indivisible finite physical atoms or that matter is infinitely divisible (without atoms), and again denies both. Transcendental idealism (a purely critical doctrine) escapes the *reductio* because, for it, the world is neither finite nor infinite in extent, and there is no infinite division of matter, nor are there material atoms. In both antinomies, the idealist has only continuing series: a series of ever more distant regions, a telescoping series of ever-finer divisions. This, we may say, is the first 'critical' use of logic.

Part I below summarizes old work of mine using intuitionistic Kripke models to explicate that critical use of logic and to defuse some venerable attacks on Kant's text. I highlight a striking technical analogy between an intuitionistic Kripke model and the structure of Kant's idealist picture. I also connect Kant's arguments to Kripke's 'assertabilist' heuristics for his intuitionistic model theory.[1]

But, of course, one must question the propriety of applying assertabilism to save an eighteenth-century text. Moreover, Kant's assertabilism comes in two forms, both of which seem ontologically off-kilter, and one of which actually goes with classical, not intuitionistic, logic.

Worse yet, my old interpretation does nothing to solve a historical anomaly about Kant's idealist version of infinity: Kant says that the second antinomy series of divisions goes on '*ad infinitum*', while the first antinomy's series of further regions continues only '*ad indefinitum*'. That's odd. For, this should be an idealist's distinction, not available to the realist. Yet, the distinction between the infinite and indefinite goes back at least to Descartes and Leibniz – realists both. As I said my original application does nothing to address this anomaly. In fact, it magnifies it.

Part II goes historical in order to solve the problems about assertability and anachronism. I outline the Descartes-Leibniz dialectic about the infinite/indefinite distinction and demonstrate how each position reflects larger philosophical positions. I then demonstrate how Kant's two versions of assertability and his version of the infinite/indefinite dichotomy emerge naturally from his attempts (first pre-critical, then critical) to resolve a tension in Leibniz's philosophy.

In light of this demonstration, Part III adapts the Kripke-model interpretation to represent that dichotomy. I then refine that formal semantics in order to depict Kant's signature critical doctrines. That, in turn, shows why Kant's critical version of the infinite/indefinite distinction is denied to his pre-critical predecessors.

Part IV turns the tables. I use what we have learned about Kant and modern semantics to address a pair of issues in the contemporary critical foundations of logic.

1. Kripke models and the antinomies

First a brief precis of basic Kripke semantics, then a rapid capsule of Kant's Antinomy.

1.1. Kripke models

Recall that intuitionistic logic invalidates excluded middle and whatever follows from it. Kripke (1965) provided a formal semantics for first-order intuitionistic logic. Given a first-order language, L, an intuitionistic Kripke model, M, for L is an ordered triple <G, K, R> (the frame) together with a domain U, a domain function ψ, and an evaluation function φ. K is the set of nodes, G ($\in$K) is the base node (the 'actual situation'), and R is the (transitive and reflexive) accessibility relation on K. Each node, H, has a domain, $\Psi(H)$ ($\subseteq$U), with the condition that HRH' implies $\psi(H) \subseteq \psi(H')$. At each node, H, the interpretation function, φ, gives the extension at H, $\varphi(H, P^n)$, for each n-ary predicate letter, P^n of L. Once again we require that HRH' implies that $\varphi(H, Pn) \subseteq \varphi(H', Pn)$.

Some propositionally compound formulae hold at a node H (we say 'are forced at H') in virtue of the local situation at that node. Thus

(1) $\varphi(H, A\&B) = 1$ iff $\varphi(H,A) = \varphi(H,B) = 1$
(2) $\varphi(H, A\lor B) = 1$ iff $\varphi(H, A) = 1$ or $\varphi(H, B) = 1$.

But the others require looking at the collection of accessible nodes:

(1) $\varphi(H, A\to B) = 1$ iff for all H'($\in$K) such that HRH', $\varphi(H',A)= 0$ or $\varphi(H',B)= 1$
(2) $\varphi(H, \sim A) = 1$ iff for all H'($\in$K) such that HRH', $\varphi(H', A) = 0$.

There's a similar asymmetry between the conditions for '$\exists$' and '$\forall$':

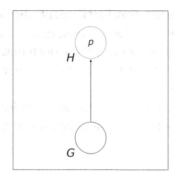

Figure 1. The Model M_1.

(1) $\varphi(H, \exists y A(x_1, \ldots, x_n, y)) = 1$ iff there is a $c \in \psi(H)$ such that $\varphi(H, A(x_1, \ldots, x_n, c)) = 1$.

(2) $\varphi(H, \forall y A(x_1, \ldots, x_n, y)) = 1$ iff for each $H'(\in K)$ such that HRH', $\varphi(H', A(x_1, \ldots, x_n, c)) = 1$ for every element, c, of $\psi(H')$.[2]

A formula *holds in the model* if it is forced at G. Kripke and others have shown that intuitionistic logic is complete with respect to this semantics.

Here is a simple Kripke model (M_1) that invalidates the law of excluded middle, but models its double negation (Figure 1).

You need multiple nodes. Single-noded Kripke models always force excluded middle and, with it, classical logic.

1.2. The Antinomy

As I said, Kant maneuvers his rival realist into a logical contradiction that he (Kant, the critical or transcendental idealist) escapes. Most famous is the first antinomy, where the realist claims that either the material universe extends finitely into space or that it extends infinitely, and then must admit that it can do neither; while Kant's idealist denies the disjunction and says that there is only an ever-expanding series of matter-occupied regions.

However, I'll start with the 'Second Antinomy', concerning the fine-grained analysis of matter. The dogmatist, Kant says, claims that any physical object is either decomposable into indivisible atoms (the Thesis) or else that object has no simple parts and is the sum of an infinite series of ever-finer parts (the Antithesis).[3] Kant then shows that for the realist neither alternative is possible, thus reducing realism *ad absurdum*. The transcendental idealist (Kant himself) escapes this contradiction. For, idealism rejects the disjunction: That object *neither* has atomic parts *nor* is it an infinite telescoping sum. Rather, says Kant, there is simply what he calls a 'regulative task', a 'rule' prescribing a search for ever finer parts and forbidding us to rest with any found division as the final one.[4]

This passage is much-vilified: Kant allegedly has an unclear, overly pictorial notion of infinity; and the whole business of 'regulative' thrust is an unnatural appendage offering a pragmatic, methodological solution to an ontological question. Moreover, the disjunction (finite or infinite divisibility) is a simple logical truth. How can Kant deny it? And, if it is not a logical truth, why is realism – again, a metaphysical, not a logical, view – committed to it? Jonathan Bennett (1974), chapters 7–9, approvingly gathered together these and other accusations. But Kripke models come to the rescue.

Let the variables of L range over components of some piece of matter, say a plank of wood (components that arise from analyzing it into finer and finer parts), and let L have one binary predicate, D(x,y), saying that x and y are such parts and y is strictly finer (comes later in the analysis) than x. Now we formalize the thesis claim about atomic parts by:

$$\exists y \forall x_{\neq y} D(x, y) \qquad (1)$$

This says that there is an ultimate legitimate division – and hence an indivisible (simple) part. The Antithesis claim about infinite divisibility becomes:

$$\forall x \exists y D(x, y) \qquad (2)$$

It says that the telescoping series of parts of has infinitely many elements.[5]

Allow these formulae as fair renditions of Kant's alternatives, and there is no problem about expressing infinity. It is Formula (2).[6]

And the logical puzzlement dissolves if we associate transcendental realism with classical logic and Kant's transcendental idealism with something like intuitionistic logic: Add the axiom that D is a strict order, and the disjunction (Thesis ∨ Antithesis) becomes a classical logical truth, but not an intuitionistic one. Here is a Kripke model (M_2) that invalidates that disjunction[7] (Figure 2).

In M_2, numbers will stand in for the plank's components. So the domain of the nth node, H_n, is {0, 1, 2, ... , n}, representing the divisions done up to that node. D (j, k) is forced at a node when j and k are in the domain of that

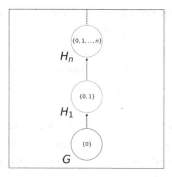

Figure 2. The Model M_2.

node and j<k. Since no node contains an absolutely smallest division, the clause for negations forces formula (3) at the base node.

$$\sim \exists y \forall x_{\neq y} D(x, y) \tag{3}$$

Similarly, since in no node does each element of the domain have a subsequent element in that domain, formula (4) holds at the base node.

$$\sim \forall x \exists y D(x, y) \tag{4}$$

Notice that this model does validate the formula

$$\forall x \sim\sim \exists y D(x, y) \tag{5}$$

Given any x and any node H_N, you cannot negate the claim that there is a division finer than x. Some subsequent node will contain such a division. This says formally – but more importantly our Kripke model depicts graphically – exactly what Kant means by the *regulative* push: We can never rest at a stage knowing that there will be no finer division; but we cannot say we have such a division until we do have it. No artificial appendage here. It stems right from the model.

So intuitionistic logic and its Kripke models nicely interpret Kant's claims about infinite divisibility. But we still need to understand Kant's transcendental idealism in a way that justifies linking it with intuitionistic logic. Transcendental idealism must serve as what Dummett calls a 'philosophical basis' for intuitionistic logic.

As I mentioned, Kant himself goes ontological here: Transcendental idealism, he says, claims that objects are 'nothing outside our representations'. Transcendental realism takes objects to be (mind-independent) things-in-themselves.[8] And the *reductio* of realism is explicitly ontological:

If the world is a whole existing in itself, it is either finite or infinite. But both alternatives are false .... It is therefore also false that the world (the sum of all appearances) is a whole existing in itself. From this it then follows that appearances in general are nothing outside our representations – which is just what is meant by their transcendental ideality. [A534-5/B506-7]

This sounds like familiar phenomenalism and realism, a fairly standard interpretation of Kant. But a wrong one: It makes Kant sound like a Berkeleyan idealist, something Kant himself denied vehemently.[9] Also, it leaves no room for objects too small or too far or too faint to be sensed. Yet Kant clearly thinks there are such objects.[10] It's logically wrong too: Ontological idealism such as this does not generate intuitionistic logic. It invalidates such intuitionistic tautologies as $\sim\sim(p \vee \sim p)$.[11]

Indeed, Michael Dummett prefers a non-ontological, *assertabilist* semantics for intuitionistic logic. In this semantics, knowledge (i.e. warranted assertability) stands in for the standard reference-based notion of truth.[12] This gives recursive semantic conditions. For instance: To assert A∨B you must be able to assert A or assert B; to assert ∼A you must show that A is never assertable; to assert ∃yA(y) you must be able to find a c such that A(c) is assertable; and to assert ∀y (Ay) you must have a method converting a proof that c is in the domain to a proof of A(c). Since we often lack the needed proofs and methods, this semantics invalidates bivalence and with it excluded middle.[13]

Certainly, there are good reasons to put Kripke models and assertabilism together. Kripke, in his original presentation, speaks of each node as representing an 'evidential situation', and describes accessibility as modeling the growth of knowledge. Quite specifically, the domain at a node H is to be the set of things that are *known* at H to exist, and Kripke's recursion conditions mimic the standard assertability conditions.

And as for Kant, the *Critique of Pure Reason* is replete with assertabilist thinking (1933). For instance there is the appropriate asymmetry between universality and existence: Universals ('All A's are B's') are true in virtue of a conceptual rule showing that any recognized A is a B; but an existential ('There exists a B') is true only when we can tie such a B to a *current* perception.[14] Recall, our model M_2 actually exploits this. And in the Antinomy, specifically, the realist supports the antithesis (the claim of infinite divisibility) by saying 'the existence of an absolutely simple cannot be established by any experience or perception, either outer or inner'. (A437/B465). This reasons from the impossibility of human experience to outright falsity. Realism here is assertabilism with classical logic.

Thus, we have a neat little package of assertabilism as the common denominator linking Kripke models to Kant.

1.3. Questions

But this exercise raises as many questions as it answers. There's the historical anomaly I mentioned: Kant says that the second antinomy's series of divisions continues *ad infinitum* while the first antinomy's series of matter-occupied spatial regions goes on only *ad indefinitum*. Kant's notion of continuability is reserved for the idealist; yet Descartes already deployed the infinite/indefinite distinction as did Leibniz, realists both. (ii) Worse yet, our model M_2 serves equally well for the first antinomy, where Kant says that the physical world extends neither finitely nor infinitely into space. Suppose that L's variables now range over parts of physical space (matter-populated regions) and that the predicate D(x,y) means that the region y is farther away from the earth than x. Then formulas (3), (4) still hold, and formula (5) still expresses the idea of a regulative pull to explore farther.[15] So this difference between the series of spatial regions and the series of material divisions seems simply to go lost.

And there are questions about Kant's idealism and the rival realism: (i) if transcendental idealism is assertabilism, why does Kant paint it ontologically? Heyting and Dummett both insist that assertabilism is ontologically neutral.[16] Moreover, why must a realist who holds that empirical objects are things in themselves also hold that the world is a thing in itself? Realism about ordinary empirical things coupled with distrust of abstracta like the sum of all empirical things is a respectable ontological position. (ii) And how can we put assertabilism in the realist's mouth? Or more precisely, why should Kant's critical philosophy favor intuitionistic logic while saddling his non-critical forerunners with classical logic?[17] (iii) Finally, how can we attribute intuitionistic logic and assertabilism to Kant at all? He did not read Dewey, Brouwer, Heyting, Dummett, or Kripke.

But he did read and internalize Descartes and Leibniz. To answer the questions about assertabilism and logic, in Part II I'll trace the dialectic about the infinite/indefinite distinction among Descartes, Leibniz and the two Kants (pre-critical and critical). I will show that each philosopher embedded the distinction in his overall view of knowledge, objects and ultimately truth. Part III will return to Kripke models. I'll use them to explicate Kant's central critical doctrines and thus to answer our questions about his critical philosophy and its version of the infinite/indefinite distinction.

2. The infinite and the indefinite: an early modern dialectic

2.1. The run-up to Kant

2.1.1. Descartes

Descartes introduced the infinite/indefinite distinction. He did it to contrast our grasp of the notion of God – unquestionably an infinite being – with the way we grasp other seemingly infinite, or at least unbounded, things: the divisions of a plank of wood for instance, or the number of stars in the firmament.[18] Our grasp of those other things – the indefinites – rests on an iterative process of extending our knowledge and on the fact that we can conjure that process's limits only in a negative, allegorical way. By contrast, we grasp God immediately, clearly and distinctly as intrinsically unlimited.[19] This brings in epistemology, ontology and basic semantics.

Epistemology: Descartes' notion of constructing geometric curves gives his cleanest epistemology of iterative processes. Here's how it works: Geometric curves

> ... can be conceived of as described by a continuous motion or by several successive motions, each motion being completely determined by those which precede; ... [I]n this way an exact knowledge of the magnitude of each is always obtainable. (Descartes 1954, *Geometrie*, Book I)

There's a process within a process: The inner process traces a continuous motion. The outer process determines the successive motions in an *effective* way: 'each motion being completely determined by those which precede'. This outer process is designed to preserve the clarity and distinctness of the inner steps. And Descartes' *Geometrie* explicitly tells us which inner motion can be clearly and distinctly grasped: Not the motions generating such curves as the quadratrix and the spiral for instance. These are approximable, but they cannot be traced by graspable continuous motions.[20] On these grounds, Descartes banishes such curves as merely 'mechanical'. The curves that can be clearly and distinctly grasped can be traced in a single continuous motion that presents all the curve's points.[21]

All the points! Grasping the whole is a criterion for admissibility, and an effective procedure must preserve that grasp. Star counting and plank dividing are not effective procedures – the next step is not guaranteed or may require choice – and carrying them out fails to provide a grasp of the complete firmament or all the plank components. That's what their 'unboundedness' comes to. The bottom line: True clear and distinct knowledge grasps the whole of its object and comes from an effective process of reasoning applied to whole-grasping, clear and distinct input.

Mathematics is the paradigm, but it is only an instance of Descartes' general 'Method', which applies across the board.

Ontology: The *Geometrie*'s lines and curves are forms of spatially extended things; and geometric constructions are abstract forms of mechanical interactions. For Descartes, the essence of an extended thing is to be physically measurable and to interact mechanically with other such things.

Semantics: Clear and distinct grasp is veridical, pure and simple. Grasp an object clearly and distinctly, it exists (as grasped). Grasp a situation clearly and distinctly, it obtains. Period.[22]

2.1.2. Leibniz

Leibniz fashioned those same Cartesian epistemological themes – clear and distinct representation, grasp of a whole, even epistemic effective procedures – into a tight-knit pristine metaphysical system. So pristine, however, that he needed special strategies to connect it with ordinary empirical objects and to preserve our empirical knowledge.

Pristine Epistemology: A *clear* grasp reveals enough properties 'to recognize the thing represented. My grasp is *distinct* if it serves to tell that thing apart from any real or possible other thing.[23] Good. But the paradigm of clear and distinct cognition is God's grasp of what Leibniz calls a 'complete concept', and that is an infinite thing. The concept of Julius Caesar, for instance, is an infinite array of descriptions that encodes every stage of Caesar's existence.[24] Unpacking that complete concept – the effective combinatory business of conceptual analysis – produces knowledge of particular facts about Caesar; indeed, facts about the whole world. For Caesar's relations with the Rubicon, with his barge, with every plank on that barge, and ultimately with everything and everyone else are encapsulated in that concept.

Pristine Ontology: (i) Objects: Only something described by a complete concept is a true *object*, an *individual substance*.[25] As such it will be predicatively complete: it either has or determinately lacks any given property at any given stage. That is its *true unity*. And because it mirrors all relations, it will express the entire world, and the world's causal structure.[26]

(ii) Collections: A description that is incomplete anywhere along the line is *general*: Delete any determination – however trivial – and you describe only the 'aggregate' of things associated with the different ways of filling that slot. Conversely, a composite thing never has a complete concept. That's why true objects are simple, indivisible, *monads*.

A composite thing has only a subjective, 'mental' or 'phenomenal' unity. It is not metaphysically real. We hold its parts and their arrangement together in our minds; and so its 'very being is also in a way mental, or phenomenal'.[27] To be sure, some aggregates have a seemingly natural unity, those that we unify according to a sortal concept (like person, or ship, or plank), but that is still only a subjective, phenomenal unity, metaphysically second class.

Pristine Semantics: This picture gives a rarified assertability semantics. The sentence 'Caesar crossed the Rubicon' is true in virtue of that combinatorial unpacking. Caesar's complete concept *contains* the concept of a Rubicon crossing. This is assertability; for, it is an epistemic state – grasping the unpacked concept – that determines truth.[28] But rarified: Only God could grasp an infinite complete concept; and so it is *God's* knowledge underlying truth here.

This assertability theory does generate classical (propositional) logic! For, as Leibniz says: 'Not true' and 'false' coincide. So 'not false' and 'true' will coincide also.... From these, it is also proved that every proposition is either true or false.... ('General Investigations of the Analysis of Concepts and Truth,' §3, 1966) This is the semantic bivalence that underlies the Law of Excluded Middle.

And Leibniz does connect assertability and ontology: God can assert all those properties that actually inhere in Caesar; so Caesar's individual concept fully expresses his true nature.[29] It is assertability united with reference.

So, this pristine Leibnizian system neatly, coherently combines an epistemic assertabilist framework with the ontology of objects and the world and with classical logic.

The Tension and Leibniz's Strategies: The pristine ontology scorns extended objects, bodies: They are divisible and cannot be true substances. The planks of Caesar's barge, for instance, and their arrangement into deck and keel, have only the subjective, 'phenomenal', unity imposed by the viewer. They are not real. Empirical science (studying extended bodies) and indeed mathematics (studying the forms of extension) are thus tainted too.

The pristine epistemology derides perception-based human knowledge. For, perception is always partial, incomplete. Stare as I might at Caesar's barge, my perception reveals neither its full past nor its future. It will not distinguish the perceived barge from others with different pasts or futures. For Leibniz, my perception is neither clear nor distinct, but rather a 'confused' conception of the barge, low on the spectrum of cognitive grasp.

And the semantics disdains empirical truth: After all, we form our empirical concepts from this confused inaccurate experience, and we form our empirical judgments from these concepts. Such judgments could not possibly accurately depict the way things truly are metaphysically.

In a nutshell: Sensory perception is confused, never veridical, and incapable of apprehending real objects.

But Leibniz was scientist and mathematician as much as metaphysician, with no wish to deride, scorn or disdain physical reality and perception-based empirical knowledge. So his philosophy is replete with strategies to reconcile the human and the pristine pulls. Here are two:

Sometimes he tried to infuse the mechanical and mathematical science of phenomena with metaphysical grounds from the pristine realm, appealing to certain 'forms of indivisible natures' as the causes of what appears rather than to the corporeal or extended mass.[30]

And sometimes, he tried a sort of *reduction*: he tried to show how empirical bodies are ultimately *composed* of monads: A physical object, he says, is repeatedly divisible, until ultimately one reaches its underlying 'monad'. 'The monads', he says, 'are the true atoms of nature; in a word, they are the elements of things'.[31]

This reduction is, I think, the key to Leibniz's own view of the infinite/indefinite distinction. That wooden plank, to be sure, has only a phenomenal unity. But it is a 'well-founded phenomenon';[32] it rests on an underlying monad whose nature dictates and unifies the stages of any division. Thus one can – God can – divide it infinitely down to that monad.[33] But the firmament has no such primary monad. In tracing *its* extent, we are left to our own devices; and Leibniz agrees with Descartes that for us that extent is at best indeterminate.[34]

2.2. Kantian strategies

Note that those Leibnizian strategies are both God-dependent: God, only God, can grasp those 'metaphysical forms of individual natures' in the infusion strategy. God, again only God, can analyze that plank down to its founding monad. Much the same holds for Leibniz's other strategies. Kant admired Leibniz's systematic metaphysics; but for Kant and many of his contemporaries, divine knowledge was an unacceptable metaphysical ploy.[35] So the Leibnizian tension remained. Kant's attempts to reconcile the empirical and metaphysical poles was a central theme throughout his philosophy, and efforts at reconciliation particularly shaped his evolving views about material division and about the extent of the firmament.

Material division was the topic of his 'Physical Monadology' of 1756, where he proposed quasi-punctual 'physical' monads that fill a space in virtue of their repulsive forces. He took up the issue of the extent of the universe in his final pre-critical work, the *Inaugural Dissertation* of 1770. There he adopted an elaborate avatar of Leibniz's pristine philosophy in a valiant attempt to reconcile the human and pristine poles. I'll look briefly at the *Dissertation*'s reconciliation strategy and its treatment of spatial extent. Then I will tell you about the *Critique of Pure Reason*'s full-blown system and Kant's treatment there of both division and extension.

The *Inaugural Dissertation*: Leibniz's opposing metaphysical *pulls* are now opposing cognitive *poles*. Kant posits two faculties: on the one hand a *Sensory Faculty* (passively received and finitely limited perceptions together with perceptually derived concepts like those of barge, river and motion) and an active *Intellectual Faculty* (with abstract concepts) on the other; both within a single human knower.

He elevates the sensible faculty epistemologically: Sensory perceptions now are 'clear and distinct' intuitions.[36] Semantically too: Empirical predications like 'Caesar crossed the Rubicon' are assertabilistically *true* conceptual containments between sensory-based concepts.[37] But ontologically it still limps: Sensory perceptions do not present authentic objects; and the sensory faculty cannot ground existence. So we have sensible assertability without reference. Indeed, *existence* and other metaphysical notions (*substance, necessity, causality*) belong strictly to the intellectual faculty. Metaphysically adequate knowledge and truth belong to the intellect.[38] So sensory perception, though clear and distinct, is formally but not metaphysically veridical.

As for the firmament: On the one hand, there's no finite end to it; it must be an infinite totality, and thus empirically ungraspable. But Kant offers a Leibniz-style infusion from the intellect to the sensible faculty. With sensibility alone we cannot achieve infinite measurements; we cannot reach the edges of the world. But that's o.k., Kant says: where our sensibility only approximates, our intellect completes. We cannot grasp the world-whole sensibly (intuitively), but no worry, the intellect kicks in and gives us the requisite grasp.[39]

The *Critique of Pure Reason*: Kant soon concluded that this inter-faculty cooperation fails. I'll turn to his reason for this conclusion shortly, but first I'll lay out the *Critique of Pure Reason*'s new conciliatory strategy.

The *Critique* elevates the sensory faculty to a full-service knowledge provider in its own right, the faculty of Understanding. 'Full service' because our human, sense-based knowledge enterprise has its own account of knowledge, its own robust ontology and its own referential semantics.

Epistemologically, sensory perception is now fully objective and veridical. Yes, Caesar only *sees* his barge and its planks cross the river, and yes, the sortal concepts *barge* and *plank* and *river crossing* organize his perception. Indeed, Kant allows no perception without a sortal concept, but now such conceptual organization is objective.[40] Moreover, those heretofore 'metaphysical' concepts (necessity, substance, existence, etc.) are now domesticated in the Understanding. Kant calls them 'schematized categories'. Their job is to show that all such combinations of experience and organization are indeed objectively justified.

And the empirical ontology no longer limps: Caesar's barge, its plank, the river, and even the event 'Caesar's river crossing' – spatiotemporal though they be – these are now full-fledged objects. So sensory perception (empirical intuition) is now indeed ontologically veridical. What we see is what there truly is. Thus the forms of perception, space and time, are the forms of reality. This is Kant's 'empirical realism'.

To be sure, Kant, like Leibniz, distinguishes truly unified objects from mere aggregates. Indeed, he goes fully Leibnizian: An actually existing object must be *predicatively complete!*[41] A compound thing about which we know there will be unanswerable questions – something with undeterminable predications – such a thing will be a mere aggregate. Since the sensation is received from an external source, Kant's point is that if you sensibly perceive the plank then you are assured that it is there to answer any question you may ask about it, even if your current perception leaves the questions unanswered.[42] If you merely conceptually contemplate that plank, you have no such assurance.

And the empirical semantics is robust: Caesar's judgment about his barge is *true* in virtue of his sortally organized perception. Existential judgments will be true in virtue of such perceptions; but universal judgments will not rest on actual perception. This is a humanized Leibnizian assertabilism; and it is now fully referential.[43]

Kant calls this humanized Leibnizian system the 'standpoint of the Understanding';[44] it comprises the epistemic, ontological and semantic norms that govern empirical discourse. And it connects Kant's assertabilism, ontology and logic together in a historically responsible way.

Let me point out that this standpoint shapes Kant's critical distinction between the *indefinite* extent of the material universe and the *infinite* divisibility of matter:

First the first antinomy account of the *extent of the universe*: To depict Kant's reasoning suppose we're a scientific task force seeking more and more distant matter-occupied regions of the universe and regularly reporting our findings. At the point when we report a series of n such regions we have grasped a large but complete object – the amalgam of all those regions. Perhaps we cannot survey it with the naked eye, it is nonetheless surveyable; and that is enough to count as a full-fledged object.[45] However, we cannot claim that there is an $n+1^{st}$ such region, and a yet larger object. We must wait to see whether we receive the needed perceptual evidence. Maybe we'll find such a region in time for the next report, maybe we will not. Indeed, if there is no further such region, we will wait forever and never know that. Similarly, we cannot now claim that we will find a next such region and one after that and one after that, forever. We have no 'effective procedure' taking us from the observation of one region to the observation of the next, from one large object to yet a larger one. We do not even know if there will be the next region. This is true, of course, for any n. Moreover, we cannot jettison our spatial skins to view the whole thing, nor our temporal skins to see how the search will look from the end of time.

Epistemologically, that sensory receptivity and that possibly infinite wait give empirical science unanswered questions (e.g. will there be a next matter-filled region) and sometimes unanswerable general questions (is this the last such region, will there be infinitely many).[46] We thus grasp that world-whole 'only in concept, never as a whole, in intuition'.[47] Ontologically, this means that the world-whole fails the test of complete determination; it is not an object at all. Semantically, as humanized assertabilists, we must deny both that the world-whole has an outermost region and that it extends infinitely.

Next, *material division*. Now we're exploring the components of that wooden plank. Once again Kant says that the collection of revealed subparts is not finite. His reasoning: Right now we grasp the plank, sortally, as a piece of wood. As we probe more we'll reveal its cellular components; and as we probe yet further we will get to the components of those cells, and – Kant will say – even more basic components of those components. (There might well be an amalgam of finer components as we proceed. Cells consist of more basic things.) But, Kant says, we will never come to a dead end with no possible further division. Space's mathematical divisibility guarantees that. Similarly as we go finer and finer we will never find a resolution with nothing but empty space.[48] For, that would preclude matter at grosser resolutions. Moreover, we'll also never hit a level with only uninterrupted matter. That's because, according to Kant, the 'attractive force' holding the parts of matter together works only in empty space.[49] So there's no finite stopping point. There's always a pattern of matter and interstices. For Kant matter is spongy all the way down.

On the other hand, we know that the telescoping collection of divisions is not infinite. Those patterns are sortally described –organized parts of an object of a given type – so as we go finer, we encounter a sortal and another and another all the way down.[50] Yet at each point in the process the assertion that there exists a further division will not yet be assertable. Perception, recall, requires sortals, but we must wait for (empirical and theoretical) science to provide the appropriate next sortal concept. The series of divisions is not only forever unfinished; it is forever innovative. There is, again, no effective procedure moving us forward. Thus we cannot yet say that the 'next' division is sortally organized and can never say that the series is 'organized to infinity'.[51]

So, once again the world-whole (now the sum of all divisions) will be neither finite nor infinite. And once again, it is only 'an *aggregate*', not an object, 'never whole'.[52] Indeed the series of divisions is not only forever unfinished; it is forever innovative. Once again there's no effective procedure to move us forward.

Now here's the difference between material division and spatial extent: We are uncertain at each report of spatial extent whether we have come to the end of the road with no further regions to be discovered. We proceed *ad indefinitum*. But no such uncertainty plagues the analysis of matter; there's no question of empty space forever or uninterrupted matter. The analysis goes on *ad infinitum*.

3. Kripke models and the critical turn

I want now to use Kripke models to precisify this Kantian version of the infinite/indefinite distinction and to distinguish Kant's version from that of his realist predecessors. First I'll model the Kantian distinction and then generalize the modeling to interpret the main lines of Kant's critical philosophy. That, in turn, will show why the distinction is denied to the realist.

3.1. Two Kripke models

Model M_2 depicts Kant's view about division: forever searching; yes, always finding; but still refraining to assert the existence of the next component. A node represents an evidential situation verifying those and only those levels of resolution so far documented. Yes, we expect matter for a finer

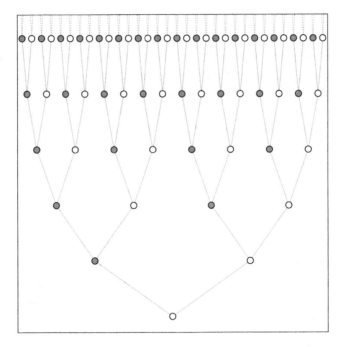

Figure 3. The Model M_3.

resolution; but *sans* a sortal concept for things that fine, we cannot assert that the level exists. The series of divisions unfinished: No nodal domain contains the full series.

But Kant's picture of matter-occupied space recommends a different model. We always face two possibilities for our next report: yes we'll encounter a further matter-filled region, no we will not. So, we contemplate Figure 3, the frame of a different Kripke model, M_3.[53]

Nodes reflect reports. Dark ones indicate finding a further matter-populated region, clear ones say no. D(x,y) says informally that the region represented by y is more distant from the earth than the one represented by x. Formally, again, natural numbers stand in for matter-occupied regions, and n>m indicates that the n-region is further away than the m-region. Each node presents an intuitable object – the region of matter-occupied space amalgamated from those regions so far discerned – and no node contains the full collection of such regions, the world-whole.

M_3 forces formulas (3) and (4), Kant's denials that the physical world extends finitely or infinitely into space. It also forces formula (5), signifying regulative pull.

Now the models' different frames precisely render Kant's distinction between the two types of series. M_3 depicts a series that continues *ad indefinitem*, where we never know whether we have reached a terminus or if we ever will. M_2 shows a series, no less regulative, no less non-infinite and non-finite, but where nevertheless at each point we confidently expect a next. This series continues *ad infinitum*.

Now to Kant's signature doctrines – Reason, synthetic *a priority*, empirical realism, transcendental idealism – and to show why the realist is denied Kant's version of the infinite/indefinite dichotomy.

3.2. The standpoint of reason

This is the standpoint of mathematics and Kant's 'transcendental philosophy'.

Mathematicians prove a theorem about e.g. triangles by abstracting from the accidental features of some found, drawn or even imagined triangle and concentrating on features it shares with all other triangles. At this abstract level, Kant sees no possible infinite wait or eternal indeterminacy. He says 'we can demand and expect nothing but assured answers' to every mathematical question.[54] Such epistemic optimism gives mathematics its own ontology and its own semantics and thus dictates a different standpoint from that of the Understanding.

Semantically, optimism supports single-noded Kripke models and classical logic. Ontologically, that, in turn, generates complete mathematical objects. For, recall, a single-noded Kripke model collects together the entire domain, and supports excluded middle (full determination) for every element of that collection. Together this epistemology, ontology and classical semantics constitute the 'standpoint of Reason'.

'Transcendental philosophy' examines empirical science's philosophical and semantic underpinnings and constraints. It justifies the schematized categories, and it shows that the series of extensions and divisions provide tasks and not facts. This metatheoretic enterprise also frees us from receptivity. 'The key', as Kant says 'is within us'. So, here too we have optimism and classical logic, the standpoint of Reason.

Ontologically, we actually get infinite things. Transcendentally we have an abstract representation of space that is 'given' as infinite.[55] We have an effective procedure for extending this abstract space without end: take a line and extend it in thought. So, when our variables range over abstract regions of space, we can assert formula (2).

In treating models M_2 and M_3, a modern transcendental task, we also adopt Reason's standpoint: No node contains the full set of divisions or extensions, but the models themselves do have that forbidden set: the global domain **U** that helps us define individual nodal domains. M_2 and M_3 are infinite objects. Moreover, Kripke's and most other completeness proofs use classical reasoning.

One could tinker with the meta-theory here. There were intuitionistic versions of Kripke models and completeness theorems.[56] But we want none of that. We want the meta-theory just exactly as Kripke sketched it. The content of the nodes remains intuitionistic; the model itself is viewed classically, from Reason's standpoint.

The models M_2 and M_3 force all the same sentences of L. They are empirically indistinguishable; only transcendentally do we see their different shapes. So, for Kant, the infinite/indefinite distinction properly belongs to transcendental philosophy; and we formulate it from Reason's standpoint.

3.3. Synthetic a priori

Mathematics studies the spatiotemporal forms of perceivable objects. Its judgments are synthetic, because they're tied to perception; they're known *a priori*, because those forms are ubiquitous. Transcendental principles, like causality, are similarly synthetic and *a priori*. With a tweak, the Kripke semantics expresses this synthetic *a priority*.

(A) *Analytic/Synthetic* Our empirical language, L, has no individual symbols. It cannot express such thoughts as: 'there will be third level components, and when we find them they will be finer-grained than the second level ones'. But we can expand L to a language L' that does express those thoughts.

Syntactically, we add countably many individual symbols, say, '0',' 'σ(0)', ..., 'σⁿ(0)', etc. ['σⁿ(0)' abbreviates 'σ(... (σ(α)) ...)' with n iterations of σ].

Informally: 0 represents our sortally distinguished starting object – e.g. that wooden plank – and $\sigma^n(0)$ stands for the sortally distinguished component uncovered at the nth level analysis. Formula (6) represents the claim 'The jth component is more refined than the kth.'

$$D(\sigma^k(0), \sigma^j(0)) \qquad (6)$$

Formula (7) represents the specific claim 'The n+1st component is more refined than the nth'.

$$D(\sigma^n(0), \sigma^{n+1}(0)) \qquad (7)$$

For the material division model we'll have '0' denote 0, and '$\sigma^{n+1}(0)$' denote n+1 in general. Then to see if an atomic sentence like (6) is forced at a node H, we would look at the referents of $\sigma^k(0)$ and $\sigma^j(0)$ to see if the pair <$\sigma^k(0)$, $\sigma^j(0)$> is in the extension of D at H. That will happen if j<k.

Clearly (7) is not forced at any node H before the n+1st. For, at least one of $\sigma^{n+1}(0)$ and $\sigma^n(0)$ will have no referent at such an H, and < $\sigma^{n+1}(0)$, $\sigma^n(0)$> could not be in D's extension there. This seems not quite right. Certainly, we can say such things from the start about the nth and n+1st components. Indeed, in Kantian terms (7) should count as an analytic truth.

This is the business of 'free logic': giving formal semantics for true sentences containing non-referring expressions. So we want a free intuitionistic semantics here, one that expands the conditions for elementary predication at a node as follows:

A predication, P($t_1, \ldots, t_k$), holds at a node H if

EITHER

(i) all of the terms ($t_1, \ldots, t_k$) have denotations in the domain of H and <$t_1, \ldots, t_k$> belongs to the extension of P at H

OR

(iia) There is at least one node H* such that all of ($t_1, \ldots, t_k$) do have denotations at H*

AND

(iib) For any node H' in which all of ($t_1, \ldots, t_k$) do have denotations, the n-tuple <$t_1, \ldots, t_k$> belongs to the extension of P at H'.[57]

In a free intuitionistic model M_2* with M_2's frame, case (i) says that P($t_1, \ldots, t_k$) holds *synthetically* at H; while in case (ii) it holds *analytically* throughout the model. (ii) guarantees (7) in the model M_2*.[58]

When we move to the question of the world's spatial extent, and read L as the language of matter-filled regions, then '0' will represent our current farthest matter-occupied region, and $\sigma^n(0)$ will stand for the nth subsequent region to be discovered. (7) now says that 'the n+1st matter-occupied region is more distant from us than is the nth'; and once again it will be analytically true.

(B) *A priority*: Our observations are fallible: The extent-of-matter task force might take a mirage for a new region or make a miscalculation. The division-of-matter team could err in identifying that n+1st component. Think of phlogiston and ether. Our modified Kripke semantics straightforwardly accommodates this fallibility: To show that the base node G has a fallible observation, just add a G' – inaccessible to G – reporting some incompatible observation (a different nth component, for instance). Now the sidelong, incompatible observation is itself receptive;

it too generates a constellation of projections, further observations consistent with G'. We add that constellation to the model. We might err in different ways – posit different components – so we can add more than one non-accessible constellation.

However, some things will persist – must persist – through all of these constellations. Every such constellation will have M_2's frame. In the course of material division each constellation *will have* a node with an n+1st component, an n+2nd component and so on. If we were observing an event, then each constellation would have us eventually discover the cause of that event. Similarly each constellation – each node of each constellation, in fact – will force mathematical truths. No matter what the n+1st component might be, if our language accommodated geometry, then Euclidean geometry would characterize the space in its interstices.[59]

For Kant, *a priori* propositions sometimes earn that status because they are conditions for the possibility of experience, sometimes they simply reflect the nature of human perception; but in either case, we depict this via their *a priority* by invariance in a collection of constellations.

3.4. Transcendental idealism and empirical realism

I say 'depict', not 'explain'. Nothing here explicates Kant's transcendental proofs of the *a priority* of mathematical or categorial principles. Nothing denotes his notions of transcendental object or self that figure into those proofs. On the other hand, we *can* model the restriction of those arguments to the forms of *human* intuition.

Kant prominently contemplates the possibility of finite (receptive) beings, whose forms of intuition differ from our own.[60] We depict the abstract possibility of such beings by adding new galaxies of constellations where even the principles of mathematics do not hold – one or another non-Euclidean geometry might hold – and where the categories will be differently schematized. Our expanded Kripke semantics adds additional clusters of constellations, additional galaxies, without a blink.[61] This, in fact, represents the *synthetic a priority* of mathematical claims and categorial principles: Though true throughout the constellations representing human intuition, they will not hold throughout the model, so they are not analytic.

But it is important to keep in mind that no matter how many inaccessible epicycles we might add (in which synthetic a priori principles fail), a Kripke model makes true those and only those formulas forced at its one base node. Truth is truth at the base node. This depicts Kant's empirical realism. The base node resides in the human, Euclidean galaxy. Truth 'for us' is truth, period. It is true simpliciter that our plank is wooden, and if it is square its angles sum to 360°.

This also precisely depicts Kant's transcendental idealism. Base node truths are 'ideal' because that human galaxy is one-among-many. Conformity to the human forms of intuition is, in some large metaphysical sense, contingent. That's idealism. It is 'transcendental', because in noting that contingency we are doing transcendental philosophy. Models of all shapes belong to transcendental not empirical discourse. Their frames and expansions belong to the transcendental ontology, along with the abstract notions of object, truth and self that Kant uses in his transcendental arguments. And, to repeat, when we model the transcendental discourse in which we discuss these things, we will use single-noded models. For, that discourse is governed by the standpoint of Reason.

3.5. The critical turn and transcendental realism

Mathematics impacts empirical knowledge and empirical objects. We calculate daily with arithmetic and measure with geometry. In an appropriate language, mathematical truths show up in every human node of each one of our Kripke models. Transcendental philosophy has a similar impact. In a language of events and causality, each 'human' constellation forces the eventual discovery of a cause for any perceived event. Moreover, reflection on our models (a task of transcendental philosophy) provides the regulative push for continued research into the ever larger and the ever smaller parts of the world.

Thus, one is tempted to accept the Understanding's ontology of empirically knowable objects while maintaining the epistemic optimism and classical logic that goes with mathematical and transcendental thought. Succumb to that temptation and you will admit the entire abstract domain as itself a legitimate object. Your single-noded Kripke model will collect together the entire domain into a single object and support excluded middle (full determination) for it. So you will have a modern humanized assertabilism together with classical logic and its world-whole. This is the dogmatist of the Antinomy. But this is the primrose path to antinomy: By logic this global object, the world-whole, is finite or infinite. That is, the disjunction of Formula (1) and formula (2) holds. But the Understanding's arguments still prevail; and so their negations (3) and (4) hold as well. Contradiction!

The 'critical' aspect of Kant's philosophy warns us to avoid this trap: to separate the two standpoints, and to exercise what Kant calls the 'discipline' needed to avoid the temptation of mixing them.

Now who would succumb to this temptation? Well, Kant did in the *Inaugural Dissertation*, where the intellect completes our limited empirical knowledge. The intellect's semantics is single noded, its logic classical. And so in the *Dissertation* the empirical world is not open-ended, there are no unreachable bounds or unplumbable depths. And this, Kant now thinks, is a mistake: The intellect's additions are not legitimate empirical objects; if you think they are, you fall into the Antinomy. So Kant here refutes his own prior strategy of infusion.

Leibniz, the father of reduction and infusion, is clearly in the same camp. As is Descartes. Think of his optimistic application of pure thought to all scientific questions. Reasoning alone, he says, applied to clear and distinct (ultimately intellectually derived) inputs will answer any question.[62]

All of these pre-critical positions assume that there is only one theoretical enterprise (detailing the world). Metaphysical reasoning and its objects will be part and parcel of the tool kit for that enterprise. Methodologically, Kant calls that dogmatism. Epistemologically, it is optimism. Semantically, its models are all single-noded; its logic always classical. Ontologically, it is realism: There is one domain in which the pristine and the empirical, the mathematical and the metaphysical all reside together. There is no room here for inaccessible nodes, constellations or galaxies. And that's what lets Kant reduce this amalgam *ad absurdum*.

A brief word about using these formal methods to interpret Kant. Some commentators think that Kant's pre-Fregean syntax rendered him unable to express infinity and generated his philosophical attitudes about infinite domains.[63] In fact, however, Kant is closer to modern logical syntax than is often recognized.[64] But the issue here is not *expressing* infinity, nor is it about actual versus potential infinity. All Kant's infinities are potential – even the given infinity of metaphysical space. The intuitionistic reading of the $\forall\exists$ quantifier combination in formula (2) enforces this. The question is rather how to distinguish between a fully graspable versus a forever innovative potential infinity. Kripke's intuitionistic models – whose domains grow but, in our case, remain finite– precisely capture Kant's notion of the forever innovative.[65] That is why Kant's distinction between the 'given' infinity of metaphysical space (formula 2) and the regulative infinity of divisions and expansions (formula 5) belongs only to this transcendental discourse of multi-noded models. It is denied to the single-node realist.

This also answers our question about Kant's version of the infinite/indefinite distinction. The distinction depends on different shaped Kripke models, and consequently it is denied the realist. That's what I set out to show.

4. The critical view of logic

Now let us turn the tables and use my modeling of Kant's critical philosophy briefly to address two modern philosophical questions in the foundations of logic.

4.1. Logic, ontology, and cognition

Kant's critical philosophy held that one human cognitive activity (the enterprise of empirical science) is governed by a finitary standpoint. I want to say that the Kantian notion of a standpoint addresses the issue of whether and how logic is grounded in ontology or in cognition. Contemporary philosophers have pondered this question at least since Husserl's *Logical Investigations*.

My use of formal models showed that the human standpoint generated a free-intuitionistic logic. But as we saw, logic went hand in hand with ontology, epistemology and cognition, and for that matter the pragmatics of assertion. This is what a *standpoint* is: a systematic philosophical unity in which all of these components are equal partners.

This is perhaps the deepest Leibnizian legacy. His pristine metaphysics seamlessly melded all those ingredients. I led with epistemology – clear and distinct was the head notion – but I could have started from any one of the ingredients and get the others. Kant strove to save that philosophical unity. He adjusted the ingredients of Understanding's standpoint to do so: the world cannot be an object; our grasp of it cannot be intuitive; truth must be humanly assertable. The non-classical empirical logic came with all of those together – with the cognition, the deep ontology, the pragmatics. That is its Kantian grounding.

I said that the empirical enterprise is *governed* by its standpoints. The standpoint is a set of *norms*: Norms governing what we may assert and when; what counts as knowledge and how to get it; what to expect (in fact predict) of objects; (an object must be ready to yield up all it nature and properties); how we are to reason (what inferences to accept, which to abjure). To work from the finite human standpoint is to be governed by all of these norms, that special logic among them. This, I think, is as much a source of that logic's normativity as anything else. Modern intuitionistic logic, is embedded in a full-scale finitary standpoint, a view of objects, assertion and truth.

This holds no less for Brouwer's mathematical intuitionism than for Kant's transcendental idealism. Brouwer, I believe, does mathematics from the finite human standpoint, thus joining logic with ontology and cognition.

4.2. The logic of the metalanguage

'Critical' also means that other cognitive projects – mathematics, transcendental philosophy, for instance – are governed by a different standpoint, Reason's standpoint; one that differs from the Understanding on each axis of norms.

This is not relativism. Mathematics and transcendental philosophy, for Kant, guide empirical science, but are ancillary to it; unequal partners in displaying reality. We need, in fact, the critical discipline to know when Reason's enterprises work together with science and when they must be kept apart.

In §3.2, I argued that for Kant this means that the logic of one's metatheory need not – should not – be the same as that of the object theory it serves. The Understanding and its logic govern empirical discourse, but Reason and its logic govern the metatheory. This issue of a separate meta-logic is hotly debated these days.

I daresay Brouwer did not accept Kant's view. He favored 'Understanding' all the way up. I will not here adjudicate between Kant and Brouwer. I will only say that Kant's philosophy gives a clear, coherent and 'critical' precedent for current fans of a separate meta-logic.

Notes

1. This part corresponds roughly to Posy (1992). Posy (1983) presented a preliminary version.
2. The quantifier conditions are generally done using assignments and satisfaction conditions. See Kripke (1965) or Dummett (2000) for full formal definitions of a Kripke model.
3. See A434/B462, A435/B463.
4. See A509/B537.
5. The Antithesis statement doesn't mention infinite divisibility, but in his 'Observation on the Antithesis '(A439/B467) Kant equates the Antithesis with the claim of infinite divisibility. I am using 'divisibility' and 'analysis' interchangeably. For, as Kant says, the division is 'in thought' (A434/B462).
6. To be sure Kant did not have modern formal languages, and I'll say something briefly later about the legitimacy of this formal translation.
7. I first applied Kripke semantics to the second antinomy in a talk, I gave at the conference, 'Saul Kripke, Philosophy, Language and Logic, A Celebration of his work', CUNY Graduate Center, January 25–26, 2006.
8. A491/B519.

9. See *Prolegomena*, AK IV, 375 (2002a).
10. See A226/B273.
11. Ontological phenomenalism naturally grounds non-bivalence. It leaves undetermined propositions as truth-valueless. (Otherwise it could not distinguish between an object t whose construction determines ~Qt, and one whose construction leaves Q undetermined.) But phenomenalism gives no ground for anything but minimal deviation from the classical truth tables. Thus if p is truth-valueless, so too will be ~p, ~~p, and (p∨~p). Thus ~~ (p∨~p) will be truth-valueless.
12. Dewey's (1938), I believe, is the origin of the term 'warranted assertability'. Dummett systematically takes warranted assertability as the ground for intuitionistic logic. See for instance, Dummett (1978).
13. The recursion conditions were first set out in Heyting (1930) in terms of mathematical proof. A more general version speaks simply of assertability in an epistemic situation. See Posy (1992).
14. See A225/B272.
15. Indeed, this is precisely the model I used in my early interpretations of the first antinomy. See for instance Posy (1983, 1991a, 1992).
16. See Heyting (1966) chapter 1 and Dummett (2000) chapter 7.
17. Putnam (1978, 26–27) suggests that a certain mapping between intuitionistic and classical logic unites assertabilism with classical logic. But this doesn't work for general assertabilism or for Kant. See Posy (2015).
18. Here's the *locus classicus*:

> We must not try to dispute about the infinite, but just to consider that all in which we find no limit is indefinite, such as the extension of the world, the divisibility of its parts, the number of the stars, etc. ... Thus because we cannot imagine an extension so great that we cannot at the same time conceive that there may be one yet greater, we shall say that the magnitude of possible things is indefinite. And because we cannot divide a body into parts which are so small that each part cannot be divided into others yet smaller, we shall consider that the quantity may be divided into parts whose number is indefinite. And because we cannot imagine so many stars that it is impossible for God to create more, we shall suppose the number to be indefinite, and so in other cases. ... And we shall name these things indefinite rather than infinite to reserve to God alone the name infinite, first of all because in Him alone we observe no limitation whatever, and because we are quite certain he can have none, and in the second place in regard to other things, because we do not in the same way positively understand them to be in every part unlimited, but merely negatively admit that their limits, if they exist, cannot be discovered by us. (*Principles of Philosophy*, Part I, Principles 26, 27 (AT VIIIA.15; Descartes [1970], I, 229–230))

Paul Franks pointed out to me that Ernst Cassirer attributed a similar distinction to Nicholas of Cusa (see Cassirer 1963, 14) . However the dialectic that I am tracing here clearly began with Descartes.

19. I am indebted to Anat Schechtman's reading of Descartes on these issues. See, in particular, her (2018, 27–44). However, I do not follow her on all details.
20. '[T]he *quadratrix, and similar curves*, which really do belong only to mechanics, and are not among the curves that I think should be included here, since they must be conceived of as described by two separate movements whose relation does not admit of exact determination. ' [Descartes, *Geometrie*, Book II.]

    To be sure, Descartes' properly 'geometric' curves have algebraic descriptions; but that is a consequence of the possibilities of construction, not a freestanding criterion of admissibility. See Henk Bos (2001).
21. Their algebraic expressions ultimately describe processes for producing those points.
22. See, e.g. *Meditations on First Philosophy*, Meditation II. Descartes had no inductively defined logic. The Port Royalists provided a 'Cartesian' logic.
23. Leibniz, 'Meditations on Knowledge, Truth and Ideas' (1969a).
24. See Leibniz, 'Discourse on Metaphysics', §13 (1969b).
25. Leibniz, 'Discourse on Metaphysics', §8 (1969b).
26. Leibniz, 'Discourse on Metaphysics', §9 (1969b).
27. Leibniz, *New Essays II*, ch. 2, section 7.
28. Indeed, in 'General Investigations of the Analysis of Concepts and Truth' [1686] (his attempt to give a 'formal semantics') Leibniz comes close to standard assertabilist truth conditions: A true judgement is one 'which can be proved' (§130). A false one is such that it can 'be proved that, however long an analysis is continued, it cannot be proved that it is true. ' (§57). *Universality* is just a question of unpacking a relation between a pair of general concepts, to see if one is included in the other. Existence, by contrast, is hard to prove: '[E]xistential propositions … cannot be proved except by using infinities or by a continuous analysis involving an infinite number of facts, namely, only from a complete notion of an individual, which involves infinite existence.' (§75.) See also §74.
29. This is what Leibniz means when he says: 'God … in seeing the individual notion or *haecceity* of Alexander, sees in it at the same time the basis and the reason for all the predicates which can truly be affirmed of him … ' ('Discourse on Metaphysics' §9, 1969b) Ironically, Leibniz is reverting to those very same scholastic notions that Descartes and his fellows reviled.
30. 'Discourse on Metaphysics', §18 (1969b). This is developed in his theory of 'living force' (*vis viva*). See the *Specimen Dynamicum* (1969d).
31. *Monadology*, §3 (1969c).
32. See, e.g. his letter to Remond of February 11, 1715 (G. III, 634–635, L. 658–660).
33. See Leibniz (2001, 233–235).
34. See Leibniz (1949) II, XIV.
35. … I will explain natural appearances as though they came from the constitution of nature; I cannot call upon God. For that would mean putting aside all philosophizing. (*Metaphysik Mrongovius AK* 1997, 774)
36. See *Inaugural Dissertation*, AKII, 395.
37. *Inaugural Dissertation*, AKII, 397.
38. *Inaugural Dissertation*, AKII, 395.

39. ... even if these co-ordinations [of a world-whole (CP)] could not be sensitively conceived, they would not, for that reason, cease to belong to the understanding. It is sufficient for this concept that co-ordinates should be given in some way or other, and that they should all be thought as constituting a unity. (*Inaugural Dissertation*, II, 391–392)
   See Posy (Forthcoming) for a fuller treatment of the *Inaugural Dissertation*.
40. See A103.
41. See A571-2/B599-600 and A573/B601.
42. An empirical object has no in-principle unknowable residue, no aspect which is intrinsically beyond our human ken.
43. See Posy (2000).
44. See Bxviii–xix, n.
45. Descartes and Leibniz speak also of counting stars, Kant speaks only of matter-occupied regions. For, he is concerned with making the regions into objects on their own.
46. See A480/B508. Kant is speaking here of global statements. Singular predicative questions about existing empirical objects can in principle be answered.
47. A518/B546.
48. See A514/B542.
49. See, *Metaphysical Foundations of Natural Science*, Chapter II, Proposition 7. AK 4, 512 (2002b).
50. At some point the pattern might remain the same (in a fractal way, perhaps). In this case there would be a most fundamental sortal. Kant will not allow that. Such a sortal would give a pattern characterizing matter itself. It would give additional structure and content to the notion of matter beyond its *a priori* properties. A similar observation could be made about repeating patterns. I do not know whether Kant would have considered that possibility.
51. A526/B554.
52. A525/B552.
53. The model M_3 has a more delicate formal definition than M_2. Here it is:

    $\mathbf{K} = \{ <a(0)\ldots a(k)> \mid k \in \mathbf{N}, \; a(i) \in \{0, 1\}, \; 0 \leq i \leq k\}$
    For all a, $a(0) = 0$
    $\mathbf{G} = \{a(0)\} = \{0\}$
    $<a(0)\ldots a(k)> \mathbf{R} \; <a(0)\ldots a(k), a(k+1)>$

    $\Psi(<a(1)\ldots a(k)>) = \{0, \ldots, m\}, \; m = \sum_{i=0}^{i=k} a(i)$

    $\mathbf{U} = \mathbf{N}$
    For all $<a(0)\ldots a(k)> \in K$, $\varphi(<a(0)\ldots a(k)>, F) = \{<n, m> \mid n, m \leq \sum_{i=0}^{i=k} a(i), \; n > m\}$

    That is For all $H \in K$, $\varphi(H, D(n, m)) = T$ iff $n > m$

54. A480/B508.
55. B39–40.
56. See Veldman (1976); de Swart (1977); and de Swart and Posy (1981).
57. §5.3 of Posy (2007) explores a variety of semantics for free intuitionistic logic. Posy ([1982] 1991) provides completeness proofs.
58. A Kantian nicety: The expressions σ^n (0) are *conceptual* expressions until the point (if there is one) that they come to denote. Analyticity always rests on conceptual grounds. Posy (1991b) relates this to Russell's notion of logically proper names.
59. This, by the way, depicts the difference between what Kant calls 'regulative' versus 'constitutive' *a priori* principles: The regulative ones are realized in constellations, the constitutive ones at every node.
60. See (A27/B43), B72, B138–139, B145, B149, A230-1/B283, and A286/B342.
61. The added constellations and galaxies of nodes are not formally idle appendages. The broad scope of condition (ii) for analyticity assures that they will affect the evaluation of wff's at the base node, G.
62. Kant addresses Descartes' realism in the fourth Paralogism of the *Critique of Pure Reason*. This passage explicitly works out the result of Descartes' mixing the empirical and transcendental enterprises.
63. See for instance Russell's *Principles of Mathematics*, sections 435 ff, and Friedman (1992), chapter 2.
64. In particular, Kant's philosophy underwrites the primacy of singular predication and the reality of relations, two ways in which traditional (and Leibnizian) logical syntax are pre-Fregean.
65. E.W. Beth's earlier semantics for intuitionistic logic requires a constant domain throughout a model (see Beth 1953, 1959, §145). This renders that semantics unfit for modeling Kant. Lopez-Escobar (1980, 1981) formally compare Beth and Kripke models. I will discuss the heuristic and philosophical differences between these semantic systems on another occasion.

Acknowledgements

I thank the editors for their patience and encouragement and for their helpful comments on an earlier draft. I also want to thank Saul Kripke for the work that inspired this paper and for his friendship over the years.

Disclosure statement

No potential conflict of interest was reported by the author.

Funding

Thanks to the Israel Science Foundation for its support. Early research for this essay was supported by ISF grant 914/02. Grant 1491/14 supported later research and writing.

References

Bennett, J. 1974. *Kant's Dialectic*. Cambridge: Cambridge University Press.

Beth, E. W. 1953. "Sur la descriptionde certains modèls d'un system formel." *Actes du XIeme Congres International de Philosophie* 5: 64–69.

Beth, E. W. 1959. *The Foundations of Mathematics*. Amsterdam: North Holland.

Bos, H. 2001. *Redefining Geometrical Exactness: Descartes' Transformation of the Early Modern Concept of Construction*. Springer.

Cassirer, E. 1963. *The Individual and the Cosmos in Renaissance Philosophy*. Translated by M. Domandi. Chicago, IL: University of Chicago Press.

Descartes, R. *Meditations on First Philosophy AT v. 7, HR 1*.

Descartes, R. *Principles of Philosophy AT v. 8, HR 1*.

Descartes, R. 1954. *Geometrie AT v. 6*. Translated by D. Smith and M. Latham. New York: Dover.

Descartes, R. 1970. "Principles of Philosophy." In *The Philosophical Works of Descartes*. Vol. 1., edited by E. S. Haldane and G. R. T. Ross. Cambridge: Cambridge University Press.

de Swart, H. C. M. 1977. "An Intuitionistically Plausible Interpretation of Intuitionistic Logic." *The Journal of Symbolic Logic* 42: 564–578.

de Swart, H. C. M., and C. Posy. 1981. "Validity and Quantification in Intuitionism." *Journal of Philosophical Logic* 10: 117–126.

Dewey, J. 1938. *Logic: The Theory of Inquiry*. New York: Henry Holt and Co.

Dummett, M. A. 1978. *The Philosophical Basis of Intuitionistic Logic*. Cambridge, MA: Harvard University Press. Reprinted in *Dummett's Truth and Other Enigmas*.

Dummett, M. A. 2000. *Elements of Intuitionism*. 2nd ed. Oxford: Oxford University Press.

Friedman, M. 1992. *Kant and the Formal Sciences*. Cambridge, MA: Harvard University Press.

Heyting, A. 1930. "Die formalen Regeln der intuitionistischen Logik." *Sitzungsberichte der preussischen Akademie von Wissenschaften* 42–56.

Heyting, A. 1966. *Intuitionism*. 2nd ed. Amsterdam: North Holland.

Kant, I. 1933. *Critique of Pure Reason*. Translated by N. Kemp-Smith. New York: Macmillan.

Kant, I. 1997. "Metaphysik Mrongovius." In *Kant's Lectures on Metaphysics*, edited by K. Ameriks and S. Naragon, 109–286. Cambridge: Cambridge University Press.

Kant, I. 2002a. "Prolegomena to Any Future Metaphysics That Will Be Able to Come Forward as a Science AK IV." In *Immanuel Kant: Theoretical Philosophy After 1781*, edited by H. Allison and P. Heath, 49–169. Cambridge: Cambridge University Press.

Kant, I. 2002b. "Metaphysical Foundations of Natural Science, AK IV." In *Immanuel Kant: Theoretical Philosophy After 1781*, edited by H. Allison and P. Heath, 171–270. Cambridge: Cambridge University Press.

Kripke, S. A. 1965. "Semantical Analysis of Intuitionistic Logic, I." In *Formal Systems and Recursive Functions: Proceedings of the Eighth Logic Colloquium, Oxford, July 1963*, edited by J. Crossley and M. Dummett, 92–130. Amsterdam: North Holland.

Leibniz, G. 1949. *New Essays Concerning Human Understanding*. 3rd ed. Translated by A. G. Langley. Open Court.

Leibniz, G. 1966. "General Inquiries About the Analysis of Concepts and of Truths." In *Logical Papers*, edited by G. Parkinson and G. Leibniz, 47–87. Oxford: Oxford University Press.

Leibniz, G. 1969a. "Meditations on Knowledge, Truth and Ideas." In *Philosophical Papers and Letters*. 2nd ed., edited by L. Loemker. Dordrecht : Reidel.

Leibniz, G. 1969b. *Discourse on Metaphysics* (in Loemker, op. cit.).

Leibniz, G. 1969c. *Monadology* (in Loemker, op. cit.).

Leibniz, G. 1969d. *Specimen Dynamicum* (in Loemker, op. cit.).

Leibniz, G. 2001. *The Labyrinth of the Continuum: Writings on the Continuum Problem 1672–1686*. Edited by T. W. Arthur and G. W. Leibniz. New Haven, CT: Yale University Press.

Lopez-Escobar, E. G. K. 1980. "Semantical Models for Intuitionistic Logics." *Studies in Logic and the Foundations of Mathematics* 99: 191–207.

Lopez-Escobar, E. G. K. 1981. "Equivalence between Semantics for Intuitionism, I." *Journal of Symbolic Logic* 46: 773–780.

Posy, C. J. (1982) 1991. "A Free IPC is a Natural Logic: Strong Completeness for Some Intuitionistic Free Logics." In *Philosophical Applications of Free Logic Topoi, v. 1*, edited by J. K. Lambert, 30–43. Oxford University Press.

Posy, C. J. 1983. "Dancing to the Antinomy: A Proposal for Transcendental Idealism." *American Philosophical Quarterly* 20: 81–94.

Posy, C. J. 1991a. "Mathematics as a Transcendental Science." In *Phenomenology and the Formal Sciences*, edited by Th. Seebohm, D. Follesdal, and J. N. Mohanty. Dordrecht: The Center for Advanced Research in Phenomenology. Kluwer Academic Publishers.

Posy, C. J. 1991b. "Kant and Conceptual Semantics." *Topoi* 10 (1): 67–78.

Posy, C. J. 1992. "Kant's Mathematical Realism." In *Kant's Philosophy of Mathematics: Modern Essays*, edited by C. J. Posy, 293–313. Dordrecht: Kluwer.

Posy, C. J. 2000. "Immediacy and the Birth of Reference in Kant: The Case for Space." In *Between Logic and Intuition: Essays in Honor of Charles Parsons*, edited by G. Sher and R. Tieszen, 155–185. Cambridge: Cambridge University Press.

Posy, C. J. 2007. "Free Logics." In *The Handbook of the History of Logic, Vol. 8*, edited by D. Gabbay and J. Woods, 633–680. Amsterdam: Elsevier.

Posy, C. J. 2015. "Realism, Reference and Reason: Remarks on Putnam and Kant." In *The Philosophy of Hilary Putnam, The Library of Living Philosophers v. XXXIV*, edited by R. Auxier, D. Anderson, and L. Hahn, 565–598. Chicago, IL: Open Court.

Posy, C. J. Forthcoming. "Of Griffins." In *Kant's Philosophy of Mathematics: Volume I: The Critical Philosophy and its Background*, edited by C. Posy and O. Rechter. Cambridge: Cambridge University Press.

Putnam, H. 1978. *Meaning and the Moral Sciences*. London: Routledge.

Schechtman, A. 2018. "The Ontic and the Iterative: Descartes on the Infinite and the Indefinite." In *Infinity in Early Modern Philosophy*, edited by O. Nachtomy and R. Winegar, 27–44. Cham: Springer.

Veldman, W. 1976. "An Intuitionistic Completeness Theorem for Intuitionistic Predicate Logic." *Journal of Symbolic Logic* 41: 159–166.

Husserl on Kant and the critical view of logic

Mirja Hartimo

ABSTRACT
This paper seeks to clarify Husserl's critical remarks about Kant's view of logic by comparing their respective views of logic. In his *Formal and Transcendental Logic* (1929, §100) Husserl criticizes Kant for not asking transcendental questions about formal logic, but rather ascribing an 'extraordinary apriority' to it. He thinks the reason for Kant's uncritical attitude to logic lies in Kant's view of logic as directed toward the subjective, instead of being concerned with a '"world" of ideal Objects'. Whereas for Kant, general logic is about laws of reasoning, Husserl thinks that formal logic should describe formal structures. Husserl claims that if Kant had had a more comprehensive concept of logic, he would have thought of raising critical questions about how logic is possible. This kind of criticism cannot itself use forms of judgments or syllogisms of logic, nor even the 'inferential' [schliessende] method more generally, but should be *descriptive* in nature. Husserl's transcendental phenomenology is the method for such criticism. The paper argues that this results in reflection, and possibly revision, of the logical principles with respect to the normative goals governing the investigation in question.

The sciences are dogmatic, they are the sciences which require "criticism" – and, indeed, a criticism, which they themselves are essentially incapable of effecting; and, on the other hand, that science having the unique function of effecting the criticism of all others and, at the same time, of itself is none other than phenomenology. (*Ideas I*, §62)

But for us, who are striving toward a radical logic, the attitude of Kant's transcendental philosophy toward formal logic is of particular interest (*Formal and Transcendental Logic*, 265/258)

1. Introduction

As a species of transcendental philosophy, Husserl's phenomenology shares Kant's critical attitude toward sciences. Like Kant, Husserl asks critical questions about the conditions of possibility about the sciences. Indeed, Husserl's primary complaint about Kant is that Kant was not critical enough:

> his [Kant's] system can certainly be characterized, in the general sense defined, as one of 'transcendental philosophy', although it is far from accomplishing a truly radical grounding of philosophy, the totality of all sciences. (Crisis, §27)

Husserl holds that Kant fails to extend his transcendental, 'how is X possible' questions to all sciences. Logic, in particular, is excluded from the critical questioning. The heart of Husserl's criticism of Kant in his mature *Formal and Transcendental Logic* (1929, hereafter FTL)[1] is that Kant failed to ask transcendental questions about formal logic:

> However greatly Kant surpassed his contemporaries, and however much his philosophy remains for us a source of profound stimulations, the half-way character of his advancement of a systematic transcendental philosophy is shown by the fact that, although to be sure he did not, like English empiricism, regard *formal logic* (taken as syllogistics, Kant's 'reine und allgemeine' logic) as mostly a worthless scholastic survival or, again like empiricism (with respect to the parts of logic it accepted), rob that discipline of its peculiar genuine sense by a psychologistic reinterpretation [Umdeutung] of its ideality, still he *asked no transcendental questions* about it, but rather ascribed to it an extraordinary apriority, which exalts it above such questions. (FTL 265/258)

Without referring to any exact passages in Kant, Husserl identifies Kant's view of general and pure logic as the reason why Kant excludes it from any transcendental scrutiny. Husserl's explanation is that Kant failed to see this because he failed to appreciate the significance of ideal formations (i.e. abstract objects) and, in particular, their place as the thematic sphere of logic. Consequently, Kant could not ask transcendental questions about it:

> Pure logic has as its thematic sphere ideal formations. But they would have had to be clearly seen, and definitely apprehended [bestimmt gefaßt sein], as such ideal objectivities, before transcendental questions about them and about pure logic could have been asked. (FTL 265/258)

---

[1] I will refer to FTL with page numbers so that the first number is to the German original and the second to the English translation. To refer to all editions in all languages at once, the references to Husserl's other works are to section numbers instead of page numbers.

In Husserl's view, logic is primarily concerned with formal theories about abstract objects; hence it is on a par with empirical sciences in respect of having a mind-independent subject matter. Indeed, by means of logic, he wanted to capture the common pure structures of the axiomatic theories (cf. Hartimo 2018).[2] Accordingly, for Husserl, the origin of logic is in Plato, and in geometry, rather than in Aristotle's syllogistics and *Posterior Analytics* (FTL 1/1). Being inspired also by Leibniz, Bolzano, and Lotze, Husserl held that the ideality of the formations with which logic is concerned is 'the characteristic of a separate, self-contained, "world" of ideal objects' (FTL 267/261). Husserl's contends that, had Kant realized this, he would not have restricted the domain accounted for by his transcendental question of 'how natural science is possible' to empirical objects, but would have noticed the need to account for also the world of ideal objectivities (i.e. abstract objects in contemporary parlance).

To put it briefly, in FTL Husserl holds that Kant failed to subject logic to transcendental criticism because he did not recognize the proper extent of the abstract, or what he calls 'ideal', logical realm. Whereas for Kant, general logic is about laws of reasoning, Husserl's view of formal logic includes the formal objects it is about. While it is debatable whether Kant thinks there are formal (mathematical) objects at all (cf. Parsons 2012, 43–49), Husserl's view of logic encompasses abstract structural mathematics. Whereas Kant, in Husserl's eyes, *uses* general logic in several ways to determine the scope and limits of reason, Husserl holds that logic, too, has to be subjected to transcendental criticism. This is also the course of Husserl's argument in FTL. He first displays his view of formal logic. After this, he is in a position to examine its conditions of possibility, which is what he does in the second part of the book, on transcendental logic. In the end, Husserl's conception of transcendental logic, which is very different from Kant's conception thereof, manifests the kind of critique formal logic should be subjected to. To be sure, 'critique' here means having been subjected to transcendental 'how is X possible' questions.

In what follows, I will try to clarify Husserl's view. I will first explain his criticism of Kant's general logic by elaborating on the differences in their respective conceptions of logic. In the end, I will expand on the critical view of logic Husserl thinks we should adopt. This will also serve as a useful starting point for a discussion of what a critical view of logic

---

[2] Husserl's usage of the term 'logic' is ambiguous. This is the sense in which he uses the term in this context, in other contexts he may refer to it as a theory of science or to a theory of judgments.

2. Husserl on Kant and the pure and general logic

In his charge that general logic is being directed towards the subjective in Kant, Husserl seems to be referring to Kant's view that general logic is about thinking, whereas Husserl holds that it should be about the objects of thought. In this sense, Husserl's conception of Kant's general logic seems correct: Kant writes that general logic is concerned with 'absolutely necessary rules of thinking, without which no use of the understanding takes place' (A52/B76). It is general, since 'it abstracts from all contents of the cognition of the understanding and of the difference of its objects, and has to do with nothing but the mere form of thinking' (A54/B78). It is pure, because 'it has no empirical principles, thus it draws nothing from psychology' (A54/B78). Kant's pure and general logic is a priori, pure, general, and independent of empirical psychology. Yet it is concerned with the subjective in Husserl's view, because it is about thinking, i.e. about subjective acts, rather than about objective structures. (For more details on Kant's view of logic, see Kjosavik's paper on the topic in the present issue).

Somewhat polemically, Husserl attributes the standpoint of the Aristotelian-Scholastic logic to it:

> Shall this unutterably defective logic be the model we should strive to imitate? No one will look kindly on the thought of pushing science back to the standpoint of the Aristotelian-Scholastic logic, which seems what Kant's treatment amounts to, since he himself says that logic has had the character of a closed science since the time of Aristotle. (Prolegomena, §58)

This complaint about Kant's view of logic seems to be, even in Husserl's own eyes, exaggerated. According to Kant, general logic is a 'concise and dry' science, 'as the methodical expositions of a doctrine of the elements of the understanding is bound to be' (A54/B78), as Kant famously puts it. Husserl cites this sentence and adds a remark: 'Everyone is familiar with Kant's lectures published by Jäsche, and knows to what a questionable extent they fulfil this characteristic demand' (Prolegomena §58). Similarly, in the passage already quoted above, Husserl complimented Kant for not regarding formal logic as a 'worthless scholastic survival' unlike the

British empiricists (FTL 265/258). Husserl was thus aware of the fact that Kant did not simply adopt the state of logic from the textbooks of his time.[3] In this sense, Kant's view of logic can also be said to be 'critical', but this notion of 'criticism' is obviously not the same transcendental sense of 'criticism' Husserl is concerned with. Independently of whether and to what extent Kant held that general logic really had reached its final state of constituting the pure doctrine of reason, Kant did not see the need for giving it transcendental foundations. Kant asked how pure mathematics is possible, how pure natural science is possible, and how metaphysics in general is possible, but he did not ask how pure logic is possible. Hence, in Husserl's view his critical philosophy does not extend to his view of pure and general logic.[4]

Husserl's further elaboration shows the generality on which his level of discussion of Kant's logic resides. Husserl thinks that Kant's failure to notice the need for transcendental foundations of logic also shows Kant's implicit indebtedness to Hume, who in fact 'was the first to grasp the universal *concrete problem* of transcendental philosophy' (FTL 257). But while Hume raised the transcendental problem of the constitution of the world, he failed to see the transcendental problem of the constitution of ideal objectivities and, in particular, of the logical idealities (FTL 266–267/259–260). This shows Hume's unquestioning acceptance of the relations of ideas, which Kant then adopted in his reaction against Hume (FTL 267/260):

> Hume directed his criticism to experience and the experienced world, but accepted the unassailableness of the relations of ideas (which Kant conceived as the analytic Apriori). Kant did the same with his counter-problem: He did not make his analytic Apriori a problem. (ibid.)

---

[3] There is ample evidence that Kant did not simply adopt the state of logic of his time. Even when he claims that with the help of the labors of logicians he was able to exhibit a complete table of the pure functions of the understanding, he added a critical remark, that the labors of logicians were nevertheless 'not yet quite free from defects' (Prol. §39). This kind of 'criticism,' however, does not count as transcendental criticism of logic, of the type 'how is X possible?,' which is Husserl's concern.

[4] In contemporary discussions, a similar criticism towards Kant has been raised by Sher (2016, 241–242). She is a defender of another critical view of logic: 'It is an interesting and a puzzling fact that a systematic philosophical foundation for logic has rarely been attempted. What I have in mind is a unified theoretical foundation, focused on logic itself rather than on mathematics, science, or natural language. Such a foundation would be critical and explanatory, and it would be especially concerned with the *veridicality* of logic. In connection with this concern it would critically examine the basic features of logic, the tasks logic is designed to perform, the source of truth and falsehood of logical and metalogical claims, the grounds on which logical theories should be accepted (rejected, or revised), the ways logical theories are both constrained and enabled by the human mind on the one hand and the world on the other, the relations between logic and related disciplines (e.g. mathematics), the normativity of logic, and so on' (Sher 2016, 239).

In Husserl's view, despite their differences, Hume and Kant share an analogous trust in the relations of ideas and the analytic apriori, respectively. In Husserl's view, this kind of unquestioned reliance on logic plagued also the rationalist tradition since Descartes, and it was also held among the transcendental philosophers after Kant (cf. esp. Crisis §§21, 25, 30). Husserl thinks that it led Kant to use 'a mythically, constructively inferring [schliessende] method' instead of a 'thoroughly intuitively disclosing [erschliessende] method' (Crisis, §30). In Husserl's eyes, Kant however came close to employing the latter method in the A-deduction, in which Kant attempted

> a direct grounding, one which descends to the original sources, only to break off again almost at once without arriving at the genuine problems of foundation which are to be opened up from this supposedly psychological side. (Crisis, §28)[5]

In Husserl's view, general logic, its forms of judgments and syllogisms, should not be unquestionably assumed in a critique of knowledge. A radical critique should not use inferences or logical 'deductions' either. Reminiscent of Frege's recourse to elucidations and Wittgenstein's say-show distinction, Husserl's criticism stems from his view that proper transcendental philosophy seeks to clarify logic by *describing* its intuitive sources (LI 1, §§1-7). For this reason, sciences and logic are excluded from the phenomenological method in the 'phenomenological reductions', as described in *Ideas I*. Phenomenological description should not presuppose them and use them but examine their givenness and constitution (Ideas I, §59).[6] Kant, however, in Husserl's view, does exactly this. He presupposes general logic and logical deductions in building his view of the apriori. Consequently, this inferring [schliessende] method results in 'mythical constructions' or in 'constructive concepts which resist in principle an ultimate clarification' (Crisis, §57).[7]

---

[5] In Ideas I Husserl claims similarly that in his A-deduction Kant operated entirely within transcendental phenomenology, but Kant misinterpreted that realm as psychological and thus abandoned it (Ideas I, §62).

[6] To be sure, Husserl's *phenomenological philosophy* is not restricted to the transcendental phenomenological *descriptions* but utilizes different kinds of methods, including argumentation.

[7] In Erste Philosophie [1925] Husserl is more elaborate about what he means by 'mythical constructions': 'Natürlich müssen wir von vornherein alle dem phänomenologischen Transzendentalismus, und damit dem tiefsten Sinn und Recht des Kantischen, widerstreitenden, in der schlechten Wortbedeutung 'metaphysischen' Bestandstücke der Vernunftkritik (wie die Ding-an-sich-Lehre, die Lehre vom *intellectus archetypes*, die Mythologie der transzendentalen Apperzeption oder des 'Bewußtseins überhaupt' usw.) übergehen und seinem mythischen Begriffe des Apriori den phänomenolgsich geklärten Begriff des allgemeinen Wesens und Wesesgesetzes substituieren ... ' (Husserl 1956, 235).

3. Husserl on formal logic

In terms of logic, Husserl's hero is Leibniz and his attempt at *mathesis universalis*:

> Leibniz, in short, had intuitions of genius: he foresaw the most splendid gains which logic has had to register since the time of Aristotle, the theory of probabilities and mathematical analyses of (syllogistic and non-syllogistic) arguments. The latter first matured in the second half of the nineteenth century. Through his *Combinatoria* he is also the intellectual father of the pure theory of manifolds, a discipline close to pure logic and in fact intimately one with it. (Prolegomena, §60)

With Leibniz's *mathesis universalis* as his guiding idea, Husserl's view of pure logic covers much more than syllogistics or any other theories of inference. It also includes the pure theory of pluralities [Vielheitslehre], pure theory of number, etc., and eventually theories of such theories, like Riemannian theory of manifolds, Grassmann's theory of extensions, Cantor's set theory (Prolegomena §§68, 70).

Husserl's logic encompasses all of mathematics also in his mature view, as expressed in FTL.[8] In it, logic is conceived of as a combination of a theory of judgments (what Husserl calls 'apophantic logic') and a formal ontology. Instead of the narrow concept of judgment in the traditional logic, Husserl counts among acts of judgment also acts of collecting, counting, ordering and combining mathematically (FTL 112/107; 269/262). Thus, to the pure theory of judgments belong, according to Husserl, 'not only the whole of syllogistics, so far as its essential content is concerned, but also (as we shall show) many other disciplines, namely those of formal-mathematical "analysis"' (FTL 60/55). Insofar as the objects of these acts of judgments are abstract, pure objectivities, they are also included in Husserl's concept of 'formal logic'. Indeed formal logic embraces formal ontology, which as a formal apriori theory of objects covers all of pure mathematics, including also non-apophantic mathematics. The latter includes traditional analysis, set theory, theories of combinations and permutations, cardinals or ordinals belonging to various levels, of manifolds, etc. [die traditionelle formale 'Analysis' der Mathematiker, die Mathematik der Mengen, der Kombinationen und Permutationen, der Anzahlen (der Modi des Wieviel), der Ordinalzahlen verschiedener Stufe, der Mannigfaltigkeiten] (FTL 81/77)). Husserl's more recent heroes in the development of logic are Riemann and Hilbert.

---

[8]FTL is according to Husserl's own assessment, his 'most mature' work, even if 'too concentrated' (Schuhmann 1977, 484–485).

Thanks to Hilbert's axiomatic approach, a clearly defined concept of a complete form of a theory was achieved around the turn of the century (FTL §31). Riemann's theory of manifolds in turn provides a universal, formal theory of such forms of manifolds. With it, the development of modern mathematics towards increasing abstraction culminates in 'having also gone on to view *such system-forms themselves as mathematical objects*' (FTL 97/93). Husserl's view of formal logic ultimately includes formal structures and their relations to each other.

Husserl thinks that his major achievement in FTL is that he managed to clarify the relationship between formal logic and formal mathematics (FTL 15/11). After extended analysis of this, Husserl's conclusion is that there is no other difference between mathematics and formal logic than the following: The full notion of formal logic includes the considerations of truth and hence a 'logical interest' towards the actually existing world.[9] Formal logic thus comes with a division into three levels. These are the pure logical grammar, the logic of non-contradiction, and the truth-logic. On different levels of logic, different kinds of evidence is sought: pure logical grammar aims at evidence in the most general sense [Evidenz im weitesten Sinne], logic of non-contradiction at the evidence of distinctness [Deutlichkeit], and finally truth at clarity [Klarheit], which requires an encounter with the world (for details, see Heffernan 1989, esp. §7, 144–153). These evidences are normative *ideals* for the practices. The difference between mathematics and formal logic is that they are guided by different normative ideals: Mathematics is only concerned with grammaticality and non-contradiction, while formal logic, conceived fully, includes also what Husserl calls 'logic of truth', in addition to pure logical grammar and logic of non-contradiction. Mathematics is thus concerned only with the first two kinds of evidence, whereas formal logic is concerned with all three.

The source of evidence of distinctness lies in the harmonious unity of possible experience, in which the contents of judgments have the coherence of matters in the synthetic unity of experience, also referred to as a

---

[9]Husserl writes: 'a formal mathematics, reduced to the above described purity, has its own legitimacy and that, for mathematics, there is in any case no necessity to go beyond that purity. At the same time, however, a great advance is made philosophically by the insight that such a restrictive reduction of logical mathesis (formal logic, when it has attained the completeness befitting its essence) —namely its reduction to a pure analytics of non-contradiction—is essentially its reduction to a science that has to do with nothing but apophantic senses, in respect of their own essential Apriori, and that in this manner the proper sense of 'formal mathematics', the mathematics to which every properly logical intention (that is: every intention belonging to a theory of science) remains alien—the mathematics of mathematicians—at last becomes fundamentally clarified. Here lies the sole legitimate distinction between formal logic and mere formal mathematics' (FTL, 146/140–141).

universe of possible experience, or a unitary sphere of experience (FTL §§89b). From the present post-Tarskian point of view, the source of distinctness appears to lie in the existence of a 'model' provided by such a synthetic unity of experience. The purpose of mathematics, according to Husserl, is thus to strive for non-contradictory theories and build 'models' for them to acquire distinct evidence of the harmony of the synthetic unity of experience in question.

Truth, excluded from the interests of the mathematicians, for Husserl means 'a correct critically verified judgment – verified by means of an adequation to the corresponding categorial objectivities 'themselves', as given in the evidential having of them themselves' (FTL 132/127). In other words, it means that the judgment is verified by the p experience of the intended state of affairs. The evidence of clarity thus involves an encounter with the actual world, hence going outside the formal realm to the world. This may take place in two ways, either directly, by applying the theories (as in applied mathematics, e.g. geometry or mechanics), or else through a 'transitional link', which is a judgment theory in which complex formations can be reduced to elementary judgments about objects (FTL §§83-84). In both cases, the ultimate goal of formal logic is to acquire fulfillment for the elementary judgments by means of perception of individual objects as something.

One might think that, with the inclusion of the domains of theories in his concept of formal logic, Husserl came close to formulating something like a Tarskian notion of logical consequence. However, Husserl is not at all interested in capturing the notion of validity of reasoning. He considers theories of inferences, syllogisms, and also tautologies, as separate theories within the logic of non-contradiction. He is *not* searching for one overarching theory of inference, or formulation of logical consequence. None of the theories of inference he mentions has any systematic role to play in his view of reasoning. Instead, his primary interest is in complete and exhaustive description of different kinds of structures and their relationships with each other. This characteristic of his view of logic can be usefully captured through Hintikka's ([1996] 2003) distinction between two roles that logic may have in mathematics. Whereas the *deductive* function emphasizes mathematical practice as theorem proving, the *descriptive* function of logic aims to characterize the structures exemplified by the intended models of the theory. Husserl's interest is clearly in the latter: Logic is about describing formal structures rather than about reasoning. Kant would probably not think of this as proper logic, at least, not as proper general logic, which has a deductive, rather than a descriptive, function.

Formal logic as characterized above can be converted into a normative-'technological' discipline (FTL 35/31). Husserl explained this in more detail already in his *Prolegomena* (1900):

> A little reflection will make matters clear. Every law of pure logic permits of an (inwardly evident) transformation, possible *a priori*, which allows one to read off certain propositions about inward evidence, certain conditions of inward evidence, from it. The combined principles of contradiction and excluded middle are certainly equivalents to the proposition: one and only *one* of two mutually contradictory judgements can manifest inner evidence. (Prolegomena §49)

The conversion is based on a general equivalence between the proposition 'A is true' and 'It is possible for anyone to judge A to be true in an inwardly evident manner' (Prolegomena §49). This process assumes an establishment of general propositions, so that

> with an eye to a normative standard, an idea or highest goal, certain features are mentioned whose possession guarantees conformity to that standard, or sets forth an indispensable condition of the latter. (Prolegomena §11)

Accordingly, later, after having distinguished between the highest goal of logic and that of mathematics, Husserl discusses the logical principles (the principle of contradiction and principle of exclude middle, *modus ponens*, and *modus tollens*) in FTL, first as principles of the logic of non-contradiction and then as principles of the logic of truth, hence relative to the normative ideals or highest goal of the formal theory in question. In both cases, they are given an objective and a subjective version. The objective versions are claims about ideal mathematical existence (logic of non-contradiction) or possibility of adequation (truth-logic), but taken subjectively, the principles relate to subjective performances. The subjective versions of the logical principles of the traditional logic relate to reasoning. They issue norms such as '[o]f two judgments that (immediately or mediately) contradict one another, only one can be accepted by any judger whatever in a proper or distinct unitary judging' (FTL 197/190). These norms are justified if they are derived from the objective principles. The subjective principles like these then establish laws of reasoning, hence something closer to Kant's general logic. But in contrast to Kant's view, the scope of the logical principles is not assumed to be general. As we will soon see, the scope of application of these principles should not be uncritically accepted. Rather, it is ultimately a matter of transcendental scrutiny with a view to the ideal goals of the underlying formal theory.

4. Husserl's critique of Kant

With this notion of logic in the background, Husserl's criticism of Kant in FTL should now be perfectly clear. Husserl writes:

> Kant's logic is presented as a science directed to the subjective – a science of thinking, which is nevertheless distinguished, as apriori, from the empirical psychology of thinking. But actually, according to its sense, Kant's purely formal logic concerns the ideal formations [idealen Denkgebilde] produced by thinking. And, concerning them, Kant fails to ask properly transcendental questions of the possibility of cognition. (FTL 267/260)

Even though Kant's logic is concerned with the subjective, Husserl concedes that, strictly speaking, Kant's purely formal logic also comes with intentions towards ideal formations. Kant, in speaking of forms of thinking, implicitly commits himself to some sort of ideality. Indeed, Husserl held that Kant and other proponents of 'formal' or 'pure' logic had correct intentions, 'but not rightly conceived and defined by them as regards its content and scope' (Prolegomena §3).

Why, then, does Husserl claim that it is subjective? As alluded to before, the reason for this is presumably that Kant's general logic is about rules for *thought*, like Husserl's subjective versions of the logical principles discussed above. In this respect, Husserl finds Kant's view of logic 'subjective' – it is about acts of reasoning by a subject. Logic is discussed in an inferential rather than in a descriptive role. Kant's view of general logic thus is a priori and theoretical, but it falls short in its content and scope. Kant's logic yields a subjective, albeit a priori, theory of reasoning. Hence, it, and everything constructed by means of it, including the sciences, is subjective:

> Like the intuited world of bodies, the whole world of natural science (and with it the dualistic world which can be known scientifically) is a subjective construct of our intellect; only the material of the sense-data arises from a transcendent affection by 'things in themselves'. (Crisis, §25)

Consequently, Kant managed to create 'a new sort of transcendental subjectivism which begins with Kant and changes into new forms in the systems of German idealism' (§25).

Husserl thinks that in its directedness to the subjective Kant's logic is too restricted, because the ideal formations, which logic intends to be about, are nevertheless excluded from the realm of logic. Husserl thinks that 'the ideality of the formations with which logic is concerned [should be taken] as the characteristic of a separate, self-contained, "world" of ideal objects' (FTL 267/261). Kant thus 'did not grasp the peculiar sense in which logic is ideal. Otherwise that sense would surely have given him a motive for asking transcendental questions' (FTL 267/261). As we saw above, Husserl's view of the ideality or abstractness of logic not only leads to inclusion of 'semantics' in his conception of logic. It also makes him see a role for logic in description of formal structures and their relationships to each other, as in Hilbert's view of axiomatics embedded in a more general theory of manifolds. For Husserl, logic is ideal in the sense that it has ideal content that is given by abstract structures. Here, Kant's view of formal logic is closer to the way many contemporary philosophers conceive of logic, whereas Husserl's view is a 'mathematician's' view of logic, including formal ontology as a theory of formal structures.

Husserl's mature criticism of Kant's view of logic curiously resonates with a way in which Husserl earlier thought that the pure idea of logic necessitated *categorial intuition*. Whereas for Kant, intuitions and categories belong to different faculties, so that the very notion of categorial intuition would be nonsensical for him, for Husserl the term captures the idea that categorial structures – the ideal structures that the world assumes – are outside of us, and they can be intuited.

Through categorial intuition, categorial objectivities (such as states of affairs) are given to us. At the time of discussing it (i.e. in *Logical Investigations*, 1900–1901), Husserl had not yet explicitly formulated his own philosophy as transcendental philosophy. But insofar as transcendental philosophy is a study of subjective achievements that make objective knowledge possible, categorial intuition was already a transcendental philosophical concept. The main change in Husserl's later view is that categorial intuition is further differentiated into different kinds of evidences (e.g. distinctness and clarity) with which formal sciences are given. In particular, his view of distinctness as a form of evidence related to a non-contradictory theory is novel in FTL. While categorial intuition is founded on perceived individuals, evidence of distinctness arises within the formal theory itself, without any comparison with the world.[10]

---

[10] Categorial intuition is crucial to Husserl's notion of truth discussed in *Logical Investigation* (1900-1901). It is founded on sensuous perception of immediately given objects so that in it, e.g. a states of affairs that encompasses also the ideal formation of the objects, is constituted (esp. LI 6, §§45-48). Transcendental philosophy can be viewed as a study of how the objective senses and validities are constituted, i.e. in what way they are achievements of subjectivity. Thus, categorial intuition is one such achievement, that is, constitution of e.g. states of affairs. Husserl claims to have discovered this 'apriori of correlation,' i.e. the correlation between objectivities (states of affairs) and their constitution, in 1898: '[t]he first breakthrough of this universal a priori of correlation between experienced object and manners of givenness (which occurred during work on my Logical Investigations around 1898) affected me so deeply that my

Husserl then thinks that once one has obtained the correct view of the abstractness of logic it should be clear that it has to be subjected to transcendental scrutiny. Indeed, understanding formal logic as a study of structures makes it rather obvious that, parallel to the transcendental problems of nature, it too has to be subjected to criticism:

> the *transcendental problem* that *objective logic* (taken no matter how broadly or narrowly) must raise concerning its domain of ideal objectivities takes a position *parallel to the transcendental problems of the sciences of realities*, the problems that must be raised concerning the regions of realities to which those sciences pertain, and in particular, the transcendental problems concerning nature, which were treated by Hume and Kant. It seems, then, that the immediate consequence of bringing out the world of ideas and, in particular (thanks to the effectuation of impulses received from *Leibniz, Bolzano,* and *Lotze*), the world of ideas with which pure logic is concerned, should have been an *immediate extension* [sofortige Übertragung] of transcendental problems to this sphere. (FTL 271/264)

Even with all the problems Husserl finds in Kant, he (Husserl) nevertheless thinks that Kant's philosophy is on a way to proper transcendental philosophy, 'that it is in accord with the formal, general sense of a transcendental philosophy in our definition' (Crisis §27). Indeed, he writes that, if Hume is counted out, 'the Kantian system is the first attempt, and one carried out with impressive scientific rigorousness, at a truly universal transcendental philosophy ... ' (Crisis §27). But, due to his subjective conception of logic and 'inferring' method, Kant managed to build a philosophy restricted to 'transcendental subjectivism' (Crisis §25). Thus, Kant's aim – to establish the scope and limits of reason – led him, for example, to distinguish between appearances and things in themselves, whereas Husserl finds such 'metaphysical' constructions unfounded.

5. Parsons's and Husserl's critical views of logic

Charles Parsons (2015) has introduced the term 'critical view of logic' to characterize positions that question the putative self-evidence of logic, in particular, the applicability of the usual logical laws in mathematics.

> The basic idea of the view is that we cannot take for granted the familiar logical principles and inferences in doing mathematics, in particular when our reasoning involves the infinite, even in the very low-level way in which the infinite enters into reasoning about natural numbers (2015, 2).

---

whole subsequent life-work has been dominated by the task of systematically elaborating on this a priori of correlation' (Crisis 166n).

While an obvious pioneer of the view is Brouwer, Parsons holds that the critical view could also be had independently on constructivist grounds. For Parsons, the view relates to what he calls 'entanglement of logic and mathematics', which refers to the way in which one's choice of logic is dependent on one's *mathematical* commitments (Parsons 2015, 13). The view thus is 'mathematics-first and philosophy-second' – it does not impose restrictions upon logic for philosophical reasons, but for reasons that have to do with the nature of the subject matter.

To obtain a critical view of logic, Husserl subjects formal logic, as discussed above, to a transcendental scrutiny. This takes place in transcendental logic:

> Logic must overcome its phenomenological naivete; even after having learned to recognize that which is ideal, logic must be more than a merely positive science of logico-mathematical idealities. Rather, with a continuously two-sided research (results on either side determining inquiries on the other), logic must go back systematically from the ideal formations to the consciousness that constitutes them phenomenologically; it must make these formations understandable, in respect of their sense and their limits, as essentially products of the correlative structures of productive cognitive life, and it must thereby fit them, like each and every other objectivity, to the broader, the *concrete*, nexus of transcendental subjectivity. The ideal objectivity of the formations with which logic is concerned – like the real world– is in no way altered in the process. (FTL 270/263)

Transcendental logic thus examines the way in which formal logic relates to the consciousness that has constituted it. This does not mean that formal logic is *constructed* and that Husserl's view is constructivistic.[11] Rather, to him formal logic is what it is to mathematicians, it is likewise approached from the 'mathematics first' stance. The task of transcendental philosophy is to understand and clarify how it is given to us; it is not to force it into any particular form of givenness. It thus studies the kinds of evidences in which logic is given, what kinds of presuppositions logic relies on, and how all this is connected to form a harmonious whole ultimately related to our lives.

This means that formal logic has to be examined and clarified in a transcendental phenomenological attitude. Whereas in *Ideas I* Husserl effects the required change of attitude with the epoche and the phenomenological reductions, more or less, in one blow, in FTL his path is much more gradual. Husserl starts by looking at the evidences with which different

---

[11] I explain the differences between constitution and construction in detail in Hartimo (2019).

layers of formal logic are given. These are then clarified and purified so that possible overlaps of evidences and other such confusions are revealed. He writes, 'evidence of every sort ... should be reflectively considered, reshaped, analyzed, purified, and improved; and that afterwards it can be, and ought to be, taken as an exemplary pattern, a norm' (FTL 184/176). Examination of evidences brings to the fore various kinds of presuppositions that are assumed in formal logic. These are, for example, the ideality of judgments, reiteration 'and so forth', and the aforementioned logical principles. Husserl discusses the logical principles separately for the logic of non-contradiction and for the truth-logic; hence, their validity is discussed relative to the evidences governing the discipline in question. Furthermore, in both cases, the principles are discussed in both an objective and a subjective formulation, where the latter is 'an evidential correlate' of the former. Husserl then examines the kinds of evidences presupposed in these principles. Without this kind of examination, a false range of application may be attributed to them:

> Because of the formal abstractness and naïveté of the logician's thinking, such never-formulated presuppositions can easily be overlooked; and consequently a false range can be attributed even to the fundamental concepts and principles of logic. (FTL 207/200)

A logic that accords with a critical view of logic is thus a transcendentally clarified logic. It is any logic or mathematics that is fully cognizant of its own aims and the kinds of evidence related to these. The evidences are clarified and purified, so that, e.g. distinctness and clarity are not confused with each other, the employed basic concepts are correct and used with their proper scopes, likewise the employed logical principles are in accordance with the demands of various evidences and used only where explicitly deemed valid. For Husserl, only science examined in this way is genuine science. Such a science is not plagued by paradoxes:

> Truth is that sciences that have paradoxes, that operate with fundamental concepts not produced by the work of originary clarification and criticism, are not sciences at all but, with all their ingenious performances, mere theoretical techniques. (FTL 189/181)

Logic or mathematics, for Husserl, should not be a mere (even if fun) game, but it should serve critically examined purposes with clarified concepts and principles with corrected scopes of application. This means to adopt the 'radicalness of scientific self-responsibility' [Radikalismus wissenschaftlicher Selbstverantwortung] (FTL 8/4).

Husserl's critical view of logic accords well with Parsons's formulation in so far as it demands a reflective justification of the usage for the familiar logical principles and inferences. The clear difference between the two is that for Husserl the applicability of logical principles is examined in relation to the goals and evidences that are sought in each discipline. This suggests that the applicability of logical principles should be assessed relative to the normative standards that govern the chosen methods (e.g. whether proofs should be constructive, whether definitions should be predicative, etc.), which leads to a more pluralistic approach. The approach is still moderately revisionary, since the concepts and principles of various disciplines will be evaluated. But, crucially, this criticism takes place internally, against the standards of each discipline. For Parsons, the source of critique lies in certain mathematical facts (e.g. about sizes of domains). His approach emphasizes the entanglement of mathematics and logic, and, accordingly, his level of analysis is technically much more refined than it is in Husserl. For the foundations of contemporary logic, a combination of these two views appears most promising. In this sense, a critical view of logic should include reflection on the choice and scope of normative standards that also takes into account the nature of the domain in question.[12]

6. Conclusion

On Husserl's view, formal logic is more comprehensive than Kant's general logic. It comprises all the formal sciences that aim at different kinds of evidence. Husserl's logic is about formal structures, whereas Kant's view of general logic is about laws of reasoning. This is why Husserl claims that Kant's view of general logic is directed to the subjective, whereas his own view of logic is rather directed towards the objective that is needed to justify the subjective principles. This makes a difference also with regard to the role of logic in each philosopher's critical philosophy: While general logic is a starting point and neutral tool for Kant's critical endeavor, Husserl's formal sciences are subjected to a critical reflection in what Husserl calls 'transcendental logic', which is not what Kant means by 'transcendental logic'. Consequently, for Kant categories have

---

[12]Husserl's view of logic differs from Sher's (2016) critical view of logic, in that she attempts to give logic theoretical foundations, while Husserl wants to give logic transcendental foundations that reveal the very conditions of possibility of logic. The way both approaches use the notion of invariance invites further comparison, which, however, has to be left for another occasion.

their basis in the understanding, whereas for Husserl categoriality is out there in the world or in the realm of ideal formations. While Husserl thinks that Kant fails to ask transcendental questions about logic, and hence to formulate a critical view of logic, for Husserl critical logic is to be identified with transcendentally reflected logical pluralism. It is a formal science carried out with an explicit awareness of the purposes of its diverse approaches and the constitution of the used principles and concepts with respect to the aims of the formal theory in question.

Acknowledgements

I want to thank Frode Kjosavik, Øystein Linnebo, Fredrik Westerlund, Sara Heinämaa, and Ofra Rechter for many valuable comments on earlier versions of this paper.

Disclosure statement

No potential conflict of interest was reported by the author.

References with abbreviations

FTL	Husserl, Edmund (1974). *Formale und transzendentale Logik: Versuch einer Kritik der logischen Vernunft* (1929). Husserliana XVII. Ed. P. Janssen. The Hague: Martinus Nijhoff. English translation: *Formal and Transcendental Logic*, translated by Dorion Cairns. Martinus Nijhoff. The Hague. 1969.
Ideas I	Edmund Husserl (1976). *Ideen zu einer reinen Phänomenologie und phänomenologischen Philosophie. Erstes Buch: Allgemeine Einführung in die reine Phänomenologie.* Husserliana III. Ed. K. Schuhmann. The Hague: Martinus Nijhoff. English Translation: *Ideas pertaining to a pure phenomenology and to a phenomenological philosophy*, translated by F. Kersten. Kluwer, Dordrecht, Boston, London.1982.
LI 1-6.	Edmund Husserl. *Logische Untersuchungen. Zweiter Band. Untersuchungen zur Phänomenologie und Theorie der Erkenntnis.* Husserliana XIX, Edited by Ursula Panzer. Martinus Nijhoff. The Hague, 1984. English translation *Logical Investigations*, translated by J. N. Findlay. Volumes One & Two. Humanities Press. New York, 1970.
Prolegomena	Edmund Husserl (1975). *Logische Untersuchungen. Erster Band. Prolegomena zur reinen Logik.* Husserliana XVIII. Ed. E. Holenstein. The Hague: Martinus Nijhoff. English translation in *Logical Investigations*, translated by J. N. Findlay. Volume One. Humanities Press. New York, 1970.
Crisis	Husserl, Edmund (1976). *Die Krisis der europäischen Wissenschaften und die transzendentale Phänomenologie: Eine Einleitung in die phänomenologische Philosophie* (1936). Husserliana VI. Ed. W. Biemel. The Hague: Martinus Nijhoff. English translation: *The Crisis of European Sciences and Transcendental Phenomenology. An Introduction to Phenomenological Philosophy*. Translated by David Carr. Evanston: Northwestern University Press, 1970.
A/B	Kant, Immanuel. (1998). *Critique of Pure Reason*. Edited and translated by Paul Guyer and Allen W. Wood. New York: Cambridge University Press.
Prol	Kant, Immanuel. (1977). *Prolegomena to any Future Metaphysics*. Translated by Paul Carus, revised by James W. Ellington. Indianapolis/Cambridge: Hackett Publishing Company.

References

Hartimo, Mirja. 2018. "Husserl on Completeness, Definitely." *Synthese* 195 (2018): 1509–1527.

Hartimo, Mirja. 2019 . "Constitution and Construction." In *Constructive Semantics – Meaning in Between Phenomenology and Constructivism*. Logic, Epistemology and the Unity of Science Series, edited by Christina Weiss. Springer. doi:10.1007/978-3-030-21313-8.

Heffernan, George. 1989. *Isagoge in die phänomenologische Apophantik*. Dordrecht: Kluwer Academic Publishers.

Hintikka, Jaakko. ([1996] 2003). *The Principles of Mathematics Revisited*. Cambridge: Cambridge University Press.

Husserl, Edmund. 1956. *Erste Philosophie (1923/24). Erster Teil, kritische Ideengeschichte. Husserliana Volume VII*. Edited by Rudolf Boehm. Haag: Martinus Nijhoff.

Parsons, Charles. 2012. "Arithmetic and the Categories." In *From Kant to Husserl*, 42–68. Cambridge, MA: Harvard University Press.

Parsons, Charles. 2015. "Infinity and a Critical View of Logic." *Inquiry* 58/1: 1–19.

Schuhmann, Karl. 1977. *Husserl-Chronik, Denk- und Lebensweg Edmund Husserls*. The Hague: Martinus Nijhoff.

Sher, Gila. 2016. *Epistemic Friction: An Essay on Knowledge, Truth, and Logic*. Oxford: Oxford University Press.

Logical pluralism and normativity

Teresa Kouri Kissel and Stewart Shapiro

ABSTRACT
We are logical pluralists who hold that the right logic is dependent on the domain of investigation; different logics for different mathematical theories. The purpose of this article is to explore the ramifications for our pluralism concerning normativity. Is there any normative role for logic, once we give up its universality? We discuss Florian Steingerger's "Frege and Carnap on the Normativity of Logic" (*Synthese* 94: 143–162) as a source for possible types of normativity, and then turn to our own proposal, which postulates that various logics are constitutive for thought within particular practices, but none are constitutive for thought as such.

1. Introduction

Issues concerning the normativity of logic are particularly vexed; there are a lot of moving parts. First, there is nothing but controversy over what counts as logic, or even what logical consequence is. Various definitions appeal to modality, inference rules, the meanings of the logical terms, meaning in general, formality, relevance, and norms of rationality. And there is nothing but controversy over what normativity is. So, it is no wonder that there is not much in the way of consensus over the extent to which logic is normative. In a widely cited unpublished article, available online, MacFarlane (2004) comes up with no less then thirty-two 'bridge principles' connecting logical consequence to normativity, some of which, of course, are more plausible than others.

MacFarlane's discussion seems to presuppose a kind of monism, that there is but One True relation of logical consequence. Things are bound to get even more complicated when we ponder pluralism about logic. Presumably, we'd have thirty-two potential bridge principles for each of the logics deemed to be legitimate or correct.

Let '$\models$' stand for the (or a) relation of logical consequence, which we take, for now, to be a relation between sets of sentences (or propositions), and single sentences (or propositions). One at least relatively uncontroversial bridge principle is the following[1]:

---

[1] This principle can be weakened by strengthening the antecedent to 'If one knows that $\Gamma \models \phi$'. This would make it even less controversial. And one can follow MacFarlane and formulate a relation of logical consequence in terms of argument *forms*, or sentence schemata. The bridge principle would be that if an argument form *S* is formally valid and you apprehend a given argument

If $\Gamma \vDash \phi$ then one should not believe every member of Γ while disbelieving φ.

What should a pluralist say about an argument that is valid on one legitimate logical consequence and invalid on another? Suppose, for example, that a pluralist finds both classical logic and intuitionistic logic to be legitimate, and suppose that φ follows from Γ classically, but not intuitionistically. Then what, exactly, are the normative obligations on the reasoner, according to this pluralist? Is there a kind of pluralism about normativity, where is it permissible on one kind of normativity – the intuitionistic variety – to believe the members of Γ while disbelieving φ, but where that very combination of belief and disbelief is impermissible on another – a classical normativity?[2]

Keefe (2014) and Read (2006) extend these questions into an argument against the pluralism articulated and defended by Beall and Restall (2006). The crucial feature of Beall and Restall's pluralism is that they accept the topic neutrality of logic, along with the thesis that logical consequence is formal. So, for them, if an argument is valid in *any* legitimate logic – for them classical, intuitionistic, and at least one relevance logic – then *any* argument in that form, about any subject matter, is valid. So, given our uncontroversial bridge principle, if an argument is valid on any legitimate logic, then one should not believe every premise of the argument while disbelieving he conclusion.

Each of us has separately articulated and defended a kind of pluralism for logic, at least concerning mathematics and other rigorously deductive enterprises (see Shapiro 2014; Kouri 2016), that does *not* accept the topic neutrality of logic. For simplicity, we will here restrict ourselves to a con sideration of mathematical theories. We propose that the correct logic is dependent on the domain of discourse. In particular, the correct logic will be dependent on the mathematical theory in question.

The upshot is that there are legitimate mathematical theories that in volve and use different logics. The most interesting cases are those, like Heyting arithmetic with Church's thesis, intuitionistic analysis, and smooth infinitesimal analysis, that become inconsistent, and thus trivial, if classical logic is imposed. Each of these intuitionistic theories has a theorem that is, or at least appears to be, classically logically false. Smooth infinitesimal analysis and intuitionistic analysis, for example, have theorems in the form:

---

as an instance of S (and believe that S is a valid form), then you ought to see to it that if you believe the premises of the argument, then you do not disbelieve its conclusion (24).

[2] In terms of bridge principles, the pluralist has two options. She can either formulate a single bridge principle which holds for every legitimate logic, with some open parameter, or she can formulate different bridge principles for each legitimate logic. We think the former is probably a better, more coherent, picture, but nothing here hangs on taking a stand on that. We are not particularly interested in articulating and defending individual bridge principles.

$\neg \forall x (\phi(x) \vee \neg \phi(x))$

Imposing classical logic on intuitionistic analysis and smooth infinitesimal analysis results in a trivial theory. This conflicts with the universal applicability of logic.[3]

The above Keefe–Read argument does not apply to our own brand of pluralism. The 'conflicts' between the various logics are the result of their different subject matters. So, one is not confronted with a conflict of competing norms for one and the same subject. To be sure, the universal applicability of logic is a longstanding thesis, and we do not give it up lightly. But this is not the place to defend our pluralism in detail.

One can accept the foregoing 'data' concerning the legitimacy of the various branches of mathematics, and still insist on monism, and the universal applicability of logic: the One True Logic holds for any and all subject matters. If intuitionistic analysis and smooth infinitesimal analysis are legitimate mathematical endeavors, then excluded middle is not valid in the One True Logic. If the program of inconsistent mathematics proves viable, then the inference rule of *ex falso quodlibet* is not part of Logic. If there are theories that invoke quantum logic, then the distributive principles are not part of Logic, etc.

Perhaps this particular matter is largely terminological, concerning what one wishes to call 'logic'. There is no need to quarrel with anyone who wishes to reserve phrases like 'valid' for these universally applicable inference patterns. The resulting monism is then true by definition. If an inference pattern is not applicable somewhere, then that pattern is automatically not part of Logic. Notice, however, that the Logic that this monism supports is a rather weak one – if our suggestions concerning what counts as legitimate mathematics are correct. Perhaps is it something like what classical, intuitionistic, quantum, relevant… logics all have in common.[4] Moreover, this monism makes it difficult to determine just what counts as Logic. Who knows what mathematical systems will arise in the future?

More important for present purposes, this 'terminological' rescue for monism essentially gives up on any substantial normative role for Logic. At a minimum, logic should guide deductive reasoning, in some sense of 'guide'. If one insists on the universal applicability of logic, then logic is to guide deductive reasoning in any and all subject matters. That is one of its

---

[3] Beall and Restall are aware of this, and they go so far as to 'reject certain branches of constructive mathematics' (120), namely those that are inconsistent with classical logic:

[4] In a sense, the perspective suggested here is the opposite of that of Keefe (2014) and Read (2006). They argue that because of their acceptance of the universality of logic, Beall and Restall are committed to the legitimacy of the *union* of the various legitimate logics: an argument is Valid just in case it comes out valid in at least one legitimate logic. Here we contemplate (and dismiss) a monism that insists that an argument is valid in case it comes out so in any and all deductive situations.

purposes. But if we are right concerning the legitimacy of various branches of mathematics, then an argument is valid only if it comes out valid on any and all of the various logics that hold in the various branches. Given how weak the resulting 'intersection' seems to be, we get precious little guidance on how to reason deductively (in any and all subject matters). For all we know, there may be no Valid arguments, or no non-trivially Valid arguments. And one can wonder how anyone manages to *know that* a given argument, or argument form, is universally valid without knowing, in advance, what sorts of mathematical theories will prove to be legitimate.

The purpose of this article is to explore the ramifications for our non universal pluralism concerning normativity. Is there any interesting sense in which logic is normative, given the rejection of the universal applicability of logic?

2. Frege and Carnap

Steinberger (2017) articulates a number of ways one might make sense of the normative status of logical laws, from the perspective of Gottlob Frege and Rudolf Carnap. His presentation is elegant, and most useful for present purposes. We will use this section to provide a summary of the relevant parts of Steinberger (2017).[5] That will provide a useful background for our own resolution of the issues.

2.1. Frege

First, Steinberger notes that there is a sense in which *any* theory has normative fallout, even theories from straightforwardly descriptive scientific disciplines like physics and chemistry. If a given descriptive theory is correct, then, obviously, it sets the standard for thinking *about* the subject matter. As Steinberger puts it,

> the correct laws of physics adequately describe portions of reality. As such they set standards of correctness for our thinking with respect to the physical arena; if we want to judge, believe and infer truly about the physical world, we had better align our thinking with the laws of physics (148)

And Frege (1893, xv):

> Every law stating what is the case can be conceived as prescriptive, one ought to think in accordance with it, and so, in this sense, is accordingly a law of thought. This holds for geometrical and physical laws no less for the logical.

For Frege, logic is, among other things, descriptive of the realm of Thoughts. It is a staple of Frege's anti-psychologism that Thoughts are not subjective,

---

[5] Page numbers throughout Section 2 are from Steinberger (2017) unless otherwise noted.

existing only in this or that mind. Thoughts, for Frege, are objective entities that are (more or less) expressed by sentences of natural language. The same Thoughts are grasped by different minds. For Frege, logic is concerned with – is descriptive of – certain relations among Thoughts.

So, to paraphrase Steinberger, 'if we want to judge, believe and infer truly' *about* Thoughts, 'we had better align our thinking with the laws of' Thoughts, the descriptive target of logic.

But this is not the main normative dimension of logic, even for Frege. Logic is also aimed at regulating *all* thinking (or all deductive thinking), not just thinking about the realm of Thoughts. Or at least it would be so aimed if there were but One True Logic, i.e. if monism were true.

This brings us to a second dimension in which logic is normative, what Steinberger calls the 'prescriptive' dimension. Quoting MacFarlane (2002, 35), some laws 'prescribe what one ought to do or provide a standard for the evaluation of one's conduct as good or bad'. Frege (1893, xv) makes the same distinction:

In one sense [a law] states what is, in another sense it prescribes what ought to be. It is only in the sense in which they lay down how one ought to think that the logical laws may be called laws of thought.

A standard example of prescriptive laws are those of ethics. Those laws are not attempting to describe how we actually behave. Rather, they tell us how we ought to behave. Traffic laws are also like this. They do not describe how we actually behave when driving a car, but they tell us how we ought to behave in those circumstances (in the indicated jurisdictions). For Frege, the one true logic prescribes one how one ought to think, in any and all subject matters. We would like to find that even for pluralists about logic, like the two of us, the laws of logic carry some prescriptive force.

There are (or, at least seem to be) key differences between laws of ethics and laws concerning traffic. It does not matter whether everyone in the world drives on the right- or left-hand side of the road, but it does seem good that each society settle on a side, and once that matter is settled, through appropriate legislative channels, it follows that everyone in that society ought to conform their behavior to accord with the law. This is a conventional prescription, adopted for a given purpose. In contrast, at least for many philosophers, the laws of ethics purport to prescribe for everyone, or at least all rational creatures. That matter, of course, remains controversial.

For Frege, the prescriptive dimension of logic is more like that of ethics than that of traffic control in a given society. There is, according to Frege, no variation in correct thinking from society to society, or from subject matter to subject matter. Nor is there any arbitrariness (except of course, in which word is to stand for each item in a given Thought). Thus, his monism concerning logic.

Steinberger goes on to invoke a Kantian distinction between *regulative* and *constitutive* norms. As Searle (1969) puts it, regulative norms 'regulate antecedently or independently existing forms of behavior'. Ethics and traffic laws qualify for that, since behavior/action generally and driving are instances of 'antecedently existing forms of behavior'.

According to Searle, constitutive norms

> create or define new forms of behavior. The rules of football or chess, for example, do not merely regulate playing football or chess but as it were they create the very possibility of playing such games (Searle 1969, 33–34, see also Searle 2010, 97).

Steinberger illustrates the contrast well:

> Take the case of traffic rules. While I ought to abide by the traffic rules, I can choose to ignore them. Rowdy driving in violation of the traffic code might well result in my losing my license. Yet no matter how cavalier my attitude towards traffic laws is, my activity still counts as driving. Contrast this with the rules governing the game of chess. I cannot in the same way opt out of conforming to the rules of chess *while continuing to count as playing chess*; in systematically violating the rules of chess and persisting in doing so even in the face of criticism, I forfeit my right to the title 'chess player'. Unless one appropriately acknowledges that one's moves are subject to the rules of chess, one's activity does not qualify as playing chess.

Of course, thinking, reasoning, and the like are not 'new forms of behavior'. People thought and reasoned well before there was any logical theorizing. Nevertheless, for Frege, the laws of logic are constitutive for thinking as such. One simply cannot think without a commitment to adhere to the laws of logic, the laws of Thought. As Steinberger puts it, on Frege's behalf: 'for mental acts or states of a particular agent to count as acts or states of thinking – to count as judgments, inferences or beliefs, say – the agent must be appropriately sensitive to the laws of logic' (149). Steinberger calls this 'Frege's constitutivity thesis'.

To be sure, the rules of football and chess, Searle's two examples, were first codified with respect to pre-existing forms of play. They were not 'new forms of behavior' either. Nevertheless, in those cases, it seems correct that, at least today, a given activity does not count as football or chess unless the participants, in some sense, take themselves to be bound by the rules.

Things are not completely straightforward here, even for these simple examples. For example, it seems that one can break some of the rules of chess (say by castling incorrectly), but still be playing chess. And a player may deliberately cheat while still intending to, and generally being regarded as, playing the game in question.[6] There is an interesting question

---

[6] Thanks to a referee here.

then of how many of the rules of chess a player can break before she can no longer be said to be playing chess at all. Surely, this is a vague and perhaps context-sensitive matter. Similarly, one can have minor violations of logic, even systematic violations, and still be thinking, reasoning, etc. How far can one stray – deliberately or otherwise – and still be considered to be thinking is, again, a vague and perhaps context-sensitive matter. Though this is an interesting issue, we will not pursue it further here.

To summarize, according to Steinberger, Frege has logic being threefold normative: its laws are descriptive of the laws of Thoughts, prescriptive for thinking and reasoning, and constitutive for thought as such. According to Steinberger, for Frege 'the descriptive [dimension of the] laws [of logic]… concern the most general regularities in the order of things; the prescriptive laws set standards of correctness to which our thinking is answerable, and which we can fail to live up to' (147).

Unfortunately, these characterizations of the normativity of logic are antithetical to the present pluralism. If the laws of logic are descriptive of an objective realm of Thoughts, prescriptive and, indeed, constitutive of thinking as such, and the logical terminology applies everywhere, then we ought always to think in accordance with the laws of Logic.[7] This entails a strong monism concerning logic: no matter what practice we are pursuing, no matter what theory we are developing, there is one set of laws to which we ought to conform our thinking. Otherwise, on the Fregean picture, we are not even thinking.

As noted in the previous section, we submit that the laws which are constitutive for thought as such are few and far between – if there are any at all. So, if we buy into the Frege picture *and* accept the variety of mathematical theories, there will not be much normativity at all. The few laws which are constitutive for Thought will not provide us with much guidance for deductive reasoning.

2.2. *Carnap*

Steinberger goes on to provide at least a possible direction for pluralism, as part of his examination of the Carnapian view of logical and normativity. According to Carnap (via Steinberger), logic is normative in a 'voluntarist' sense, in that we can *choose* which logical norms we will be subject to on various occasions.

For Carnap, the question of which system produces the One True Logic is an 'external pseudo-question'. As Steinberger puts it, for Carnap, 'it is

---

[7] We assume that Frege thought the logical terminology is everywhere applicable, and was thus universal in the same sense logic is. On our theories of logical pluralism, this is not the case (see Shapiro 2014; Kouri 2016), so the correct logical terminology for a domain of inquiry depends on the domain is question. Thanks to a referee for highlighting this issue.

we who choose logical systems, and we do so on pragmatic grounds: we adopt those logical systems that best serve our theoretical ends' (154).

Carnap's position is pluralistic in the sense that it allows that different linguistic frameworks can be generated with different logics, and that more than one of those linguistic frameworks may prove fruitful for this or that theoretical purpose.

This means that the goodness or badness of a logic is relativized to whatever our aim is. It also means that Carnap needs to tell a different story about normativity than Frege did, or run into the problems with universality we outlined above.

Carnap's views can be summarized by two often-quoted passages:

> Our attitude towards… requirements… [of a logic] is given by a general formulation in the *Principle of Tolerance: It is not our business to set up prohibitions but to arrive at conventions.*
>
> *In logic, there are no morals.* Everyone is at liberty to build up his own logic, i.e. his own form of language, as he wishes. All that is required of him is that, if he wishes to discuss it, he must state his methods clearly, and give syntactical rules instead of philosophical arguments. (Carnap 1937, 51–52)

What these quotes suggest is that any set of syntactic rules built up by some method will generate a legitimate logic. We, as logicians and philosophers, are not in a position to give philosophical arguments for or against such sets of syntactic rules. If we happen to construct two calculi with different logical syntax, and if both serve some legitimate theoretical purpose, then we will have generated a type of logical pluralism.

Carnap suggests that, though we cannot give philosophical arguments for or against logics, some of these logics might be more or less useful for certain purposes. We find that, for Carnap, what is important is not the logical system itself but whether the logical system suits the purpose it is being used for, and whether the logical system can be given by formal syntactical rules. We can accept or reject a given framework/logic pair depending on its efficiency as an instrument and whether it accomplished the goals it was intended to accomplish (see Carnap 1950a, 275).

The story Carnap tells about normativity is what Steinberger calls a 'Relatavized Constitutivity Thesis', which is meant to contrast with Frege's 'Constitutivity Thesis'. For Carnap, the laws of a single logic are not constitutive for thought across the board. There are no such laws (even asking about them is illegitimate). However, the laws of a particular logic might be constitutive for thought with respect to a certain theoretical end. Once we select the logic with which we will pursue that end, we are thereby subject to reason in accordance with the laws of that logic – provided, of course, that we continue to pursue that end.

This is the sense in which the normativity on this picture is voluntarist: we are only subject the logical norms we have selected to use. On the other hand, there is still a sense in which these norms are constitutive for *thought within* the given framework. We cannot think clearly and rigorously in any given context without some formal rules in place.

This way of conceiving of normativity means both that the laws of certain logics are constitutive for thought – within certain practices – and that they are still descriptive of those who have chosen to pursue those practices appropriately. In a sense, Frege has it that the laws of logic are categorical imperatives. Carnap has them as hypothetical imperatives tied to this or that theoretical purpose.

On our views, any coherent mathematical system will be associated with a legitimate logic. That will be the logic appropriate for reasoning within that system/context. This is in line with the Carnapian view as Steinberger describes it. So, we would like to (for the most part) adopt this way of thinking about normativity.

However, we cannot rest content here, as there is still a major question that bears answering. What is the relationship between a given formal linguistic framework and the informal, pre-theoretic practice that corresponds to it? What, for example, is the relationship between classical, Peano arithmetic, as explicitly articulated in logic books, and the practice of arithmetic and number theory by mathematicians and scientists? The former seems to qualify as a Carnapian linguistic framework; the latter is an ongoing theoretical enterprise. What do they have to do with each other?

More specifically, what is the relationship between the language/logic of the formal linguistic framework we choose to pursue our theoretical end and the natural language typically used to pursue that end? Without a satisfactory answer to this question, we cannot make sense of this voluntarism about norms, or how the logical norms relate to the actual practice of deducing things in the natural languages used in mathematics and science.

That question will be answered in our final section, after we further articulate the main source and nature of the norms that underlie deductive reasoning. We turn to the normative nature of the use of language, and to semantics in particular, and to the distinction between a descriptive and a prescriptive theory of an evaluative practice.

3. Meaning and normativity

According to Alberto Coffa (1993), a main item on the agenda of philosophy throughout the nineteenth century was to account for the necessity and a priority of mathematics and logic without invoking Kantian intuition. Coffa argues that the most successful line on this was what he calls the 'semantic tradition', that went through figures like Bolzano, Frege, Russell, and Hilbert, and culminated in the Vienna Circle. The key idea, at least toward the end of the 'tradition', is that the necessities in question lie in the *meanings* of the terms used in mathematics and logic. Accordingly, true mathematical propositions are *analytic*.

Although there are still varieties of neo-logicism on the philosophical market, the thesis that mathematics is analytic is no longer prominent. Nevertheless, the rough idea that logical consequence is somehow tied to the meanings of logical terms remains popular. One slogan is that an argument is valid if the premises guarantee the conclusion *in virtue of the meanings of the logical terminology*.

This view of logic is of-a-piece with the longstanding thesis, going back at least to Aristotle's *Prior Analytics*, that validity is a matter of form. If a given argument is valid, then so is any argument that has the same form. The idea here is that the logical terminology of a given argument is what marks its form. The non-logical terminology is what makes for its content. So, at a minimum, the meanings of the logical terms have something to do with the validity or invalidity of the argument – and, of course, the specific meanings of the non-logical terminology do not.

Of course, these various theses are not held universally, and they are often articulated in different, mutually incompatible ways. But they at least suggest that one look to *semantics* for the source of at least some of the normativity attached to logic.[8] Suppose, for example, that a given set of premises entail a given conclusion (in the relevant sense of entailment). Suppose also that someone believes all of the premises and disbelieves the conclusion, thus violating one of MacFarlane's 'bridge principles' (see Section 1). What has gone wrong? What sort of norms have been violated? On the views in question, the contents of the various beliefs are jointly impossible, and moreover, this impossibility is due to the meanings of the logical terms. If we assume also that the subject is a competent speaker of the language, then he knows, at least tacitly, what the logical terms mean. So, as a competent language user, he should know better; he should know that if the premises of the argument are true, then the conclusion is also true, and so he should know that he should not continue to believe the premises and disbelieve the conclusion. And, it seems, he surely should know better if, in addition, he *knows that* the argument in question is valid.

So, to understand the normativity of logic, it might help to consider the senses in which language use is normative. Matters here are vexed, but we will try to be as neutral as possible on the underlying issues. In the end, we hope to shed further light on the foregoing pluralism about logic, and the relevant sense of normativity involved with logic.

---

[8] This perspective requires a notion of logic which is inherently about language, or at least properly regimented language. This runs against those who hold that logic is not metalinguistic in this sense, and is a theory of the world (see, e.g. Williamson 2013). We do not have space to address this issue here.

Language use is normative in several different ways. For a start, there are correct and incorrect ways to spell individual words, and there are grammatical and ungrammatical sentences. Apparently, there is nothing more to determining what counts as correct, in these cases, than the way that the words and sentences are used by competent speakers of the language, the relevant linguistic community. Notably, this use includes speakers' practices of correcting each other on matters of spelling and grammar.

Burgess (1992, 2015) sees logic – classical logic in particular – in similar terms. He points out that classical mathematics is a normative endeavor, in the sense that there are standards for rigor that mathematics must meet. If a text fails to meet the standard of classical validity, at least up to an approximation, then it will probably not be accepted for publication in a professional journal. And if it is later discovered that a publication with an invalid inference has slipped through, the author will be compelled to withdraw the article, or else to correct the lapse in rigor (if possible). For Burgess, classical logic is understood as a theory of rigorous proof, in the same sense that a grammar is a theory of correct sentence construction. He writes:

> Whenever a community has a practice, the project of developing a theory of it suggests itself. When the practice is one of evaluation, a distinction must be made between descriptive and prescriptive theories thereof. The former aims to describe explicitly what the community's implicit standards have been: the theory is itself evaluated by how well it agrees with the facts of the community's practice. The latter presumes to prescribe what the community's standards ought to be: the community's practice is evaluated by how well it agrees with the norms of the theory. Logic, according to almost any conception, is a theory dealing with standards of evaluation of deduction, much as linguistics deals with standards of evaluation of utterances. The distinction between descriptive and prescriptive is familiar in the case of linguistics: no one could confuse Chomsky with Fowler. It is not less important in the case of logic. The familiar case of linguistics can help clarify a point about intuition important for logic. The data for descriptive theorizing consists of evaluations of members of the community whose evaluative practices are under investigation (e.g. 'That's not good English'). (Burgess 1992, 3)

For Burgess, then, logic is empirical, in the same sense in which much of linguistics is. In logic, the 'data' consists of evaluations of members of the mathematical community (or at least those engaged in classical mathematics) concerning whether inferences are correct. The logician comes up with a theory that is supposed to describe or codify those judgments, just as a grammarian comes up with a theory of syntax. For the case at hand – classical mathematics – Burgess plausibly argues that the correct descriptive theory is classical logic.

It will prove helpful to make a three-way distinction concerning various theories, extending our previous distinction between prescriptive and descriptive enterprises. If a theory is aimed at something that is not an evaluative practice, then, of course, the theory is descriptive. In this case, it is descriptive of a non-evaluative subject matter. Examples include physics, chemistry, perhaps number theory, geometry, etc. Let us say that theories like these have a *non-evaluative target*. As we have seen in the previous section, Frege (and Steinberger) have pointed out that such theories, if correct, do have normative fallout, concerning how one should think and talk *about* the non-evaluative subject matter. If one is to think and talk correctly about subatomic particles, for example, then one should only think and say…

If, instead, a theory is aimed at an evaluative practice, then Burgess's distinction applies. A *descriptive theory of an evaluative practice* 'aims to describe explicitly what the community's implicit standards' are. And the theory is correct if it accurately captures those implicit standards – up to thorny issues of idealization and the like–the same issues that continue to loom large for us. In contrast, a *prescriptive theory of an evaluative practice* 'presumes to prescribe what the community's standards ought to be'. The prescriptive theorist claims that the community's standards – the ones that they implicitly adopt and follow and use to evaluate each other – may be flawed, or incorrect, or whatever. The prescriptive theorist tells them what their standards ought to be.

We can see the latter contrast in various defenses of non-classical logics. It is reasonably clear that the early intuitionists, Brouwer (1913) and Heyting (1956), as well as some later intuitionists, such as Dummett (1973), are prescriptivists concerning mathematics and its logic. They are well aware that mathematicians reason classically, and that classical logic is a reasonable description of the norms upheld in the community of (classical) mathematicians. But they argue (on various grounds) that this practice is flawed, and they tell us what the norms ought to be.

A descriptive orientation to intuitionistic logic is also possible. One might argue that excluded middle, and the like, are, in fact, not upheld as linguistic norms, even in the community of classical mathematicians. Any apparent 'data' to the contrary is chalked up to non-logical principles endorsed (perhaps tacitly) by the reasoner.

Tennant (1996), for example, argues that the correct meaning of the logical terminology only sanctions intuitionistic logic. The law of excluded middle is, for the classical mathematician, 'synthetic a priori'. Tennant thus explicitly endorses the aforementioned connection between meaning and logic, and claims that classical logic fails to be descriptive of the meaning of the logical terms.

At least the early advocates of relevance logic, such as Anderson and Belnap (1975), criticized classical logic on descriptive grounds. They claimed that classical inference forms such as *ex falso quodlibet* (and disjunctive syllogism) are not among the norms implicit in contemporary deductive practice. A different kind of relevance logician might concede that the descriptive battle is lost – the indicated inferential practice is classical – and then wax prescriptive, arguing that the community in question ought to adopt a relevant logic.

Unfortunately, the pieces of this jigsaw puzzle do not fit together well, at least not so far. Burgess's main exemplar of a descriptive enterprise aimed at an evaluative practice is *grammar*. Spelling would do as well. In contrast, the semantic tradition, and its contemporary fallout, have it that logical consequence is connected to the *meanings* of the logical terminology. As the name suggests, the focus is linguistic, but it is on *semantics*, not grammar or spelling.

So, first, is semantics about an evaluative practice – does it have an evaluative target, one that is amenable to study by empirical methods? In a sense, for Frege, the answer is 'no'. For Frege, logic, if not semantics, is about relations in the realm of Thoughts. To be sure, there is an empirical question concerning which Thought is expressed by a given sentence (in a given context). And perhaps there are normative matters concerning which Thought a given sentence ought to express. But the realm of Thoughts is itself not a target for empirical study.

In contrast to this Fregean orientation, contemporary empirical semantics does have an evaluative target; it concerns the use of various words and expressions in communication. And, arguably, there are implicit standards in place for how words are to be used in the various linguistic communities, such as speakers of English.

There is a thorny question concerning just what form a semantic theory, for a fragment of a natural language, should take. There are varieties of traditional truth-conditional approaches, and there is dynamic semantics, such as discourse representation theory, file change semantics, and others. Among the community of logicians, there are model-theorists, who characterize logical terminology in terms of its truth-conditions, and there are inferentialists, who characterize logical terminology in terms of inference rules, typically introduction and elimination rules. We would do well to avoid having to adjudicate these broad methodological matters here.

It is often straightforward to determine the purpose(s) served by a given evaluative practice. Implicit norms typically further, or are 'designed' to further a particular telos. That is what they are for. In the case of grammar and spelling, the purposes include the facilitation of communication. Having norms for grammar and spelling, in a given linguistic community, helps the speakers to make themselves understood, and to minimize mis communication. Presumably, there is more then one grammar or spelling system that would serve this end. It does not matter whether a given word is spelled 'color' or 'colour', but surely it helps if there is only a single spelling (or perhaps a small group spellings) for each lexical entry.[9]

Similarly, as noted in the previous section, the traffic ordinances in a given society serve the purpose of facilitating travel by automobile. It serves this purpose if all cars were to drive on the same side of each street, and in just about all societies, there are ordinances mandating which side to drive on. But it does not matter which side of the road is chosen for this.

So, we have a kind of pluralism for grammar and spelling, and also for traffic ordinances. But this is nothing like the pluralism(s) concerning logic that we have articulated elsewhere. Or is it?

What is the telos of the implicit norms concerning meaning? One main purpose of language, of course, is communication, the exchange of information. In most situations, speakers try to say what is true. In deed, the 'data' invoked by semanticists consist of (or at least include) the spontaneous judgments of speakers concerning what they would say in various situations – what they would say is true in those situations. With dynamic semantics, the focus is on context-update, but in normal situations, this ultimately comes down to what speakers take to be true in the given context.

This observation, or assumption, is of-a-piece with the longstanding thesis that logical consequence is truth-preserving. If an argument form is valid, then it is not possible for the premises to be true and the conclusion false.[10]

As with grammar, spelling, and even traffic ordinances, it is conceivable that the norms upheld in a given community do not further the relevant telos. In this case, it is the goal of communication – of exchanging information about what is true. Consider, for example, a community that seemed to adopt a 'tonk' connective (Prior 1967). A descriptive semanticist would note this in giving an account of the meanings of the terms in the language in question, perhaps pointing out, in a footnote, that the overall system is badly flawed, indicating cases of miscommunication that result. A differenttheorist, orthesameoneinalaterwork, couldthengoprescriptive (assuming that the aforementioned descriptive account is correct), urging

---

[9] It might happen that the norms for grammar or spelling in a given community do not facilitate communication as well as they could. That would mean that, all too often, speakers fail to understand one another. In such a case, a theorist might turn prescriptive, urging the community to adopt different norms, or perhaps a different language.

[10] It is problematic to insist that validity is truth-preserving for languages that contain an unrestricted truth-predicate. Field (2006) has shown that with the exception of a few sub-structural and non-transitive logics, it is inconsistent to hold that even modus ponens preserves truth – thanks to the Curry paradox. Recall, however, that the present focus is on mathematics. Arguably, the languages used for mathematics do not, or at least need not, contain an unrestricted truth-predicate. We care about arithmetic, for example, and arithmetic truth, but not so much about truth in general. A Tarskian truth-predicate will suffice for mathematical purposes. It is, we presume, not at all controversial that, whatever validity is, a valid argument does (or at least ought to) preserve Tarskian truth. Thanks to Kevin Scharp for pressing this.

the speakers to drop that connective, and perhaps replace it with another one that works better.[11]

4. A Carnapian resolution

So where does this leave us? We saw that, according to Steinberger, Frege held that logic is *constitutive* of thinking (or at least of deductive reasoning). If one is to think or reason at all, then one is bound by the laws of logic – the one true logic that correctly describes how things are in the realm of Thoughts. Moreover, thinking, as such, demands that the subject at least tacitly understand herself as so bound.

As we also saw, with some constitutive norms, such as those for games like chess and football, one can avoid the norms by simply not engaging in the activity in question. However, short of committing suicide, one does not have the option to simply not think. We presume that there is also no serious option to avoid deduction, when thinking. So, in quasi-Kantian terms, the norms correctly described by logic, on this Fregean perspective, are categorical: they apply to any and all deductive reasoning about any and all subjects. And, of course, this is a thorough monism concerning logic (but, so far, we have no conclusions as to what the One True Logic is).

Once again, we join the Carnapian in rejecting the existence of Frege's single realm of Thoughts, and with that, we reject the existence of such 'categorical norms' for all deductive reasoning, as such (or, at least we reject the thesis that logic provides norms like this).

As we see things, the norms of various logics are constitutive of various mathematical theories: there are classical theories, intuitionistic theories, and perhaps even relevant theories, quantum theories, sub-structural theories,… And of course, any given deductive theory is, in some sense, optional: one can choose not to use it, or not to pursue it. So, all we have are hypothetical imperatives. If you are working in a classical theory, you ought to reason in such and such a way; if in an intuitionistic theory, in this and so way, etc. This much is compatible with the foregoing pluralism about logic.

But this does still not completely settle the foregoing questions concerning the normativity of logic. In the previous section, we tied the normativity of logic to the normativity of language – to the normativity of semantics in particular. Mathematicians do not generally work in the formal systems developed by logicians. In their natural habitat, they prove theorems and communicate in ordinary language, as supplemented with mathematical

---

[11] Of course, there are other, less drastic ways that a given language can fail in its telos of facilitating communication. For example, the sentences may be more complex than is necessary to convey the relevant information. Here, too, a descriptive theory, a semantics, would attempt to say what the implicit norms for meaning are, and a prescriptive theory would suggest improvements.

symbols and jargon of course. What are we to say about how one ought to reason in everyday mathematics?

Similar questions arise for the Carnapian generally, and for Carnap himself. What is the relationship between a linguistic framework – a formalized mathematical theory in a given logical system – and the mathematics that it corresponds to?

Recall that for Carnap, if 'someone wishes to speak in his language about a new kind of entities, he has to introduce a system of new ways of speaking, subject to new rules' (Carnap 1950a, 242). Recall our more or less standard quote[12]:

In logic, there are no morals. Everyone is at liberty to build up his own logic, i.e. his own form of language, as he wishes. All that is required of him is that, if he wishes to discuss it, he must state his methods clearly, and give syntactical rules instead of philosophical arguments. (Carnap 1937, 51–52)

This clearly normative pronouncement is usually taken to be aimed at traditional metaphysics, at attempts to address external pseudo-questions, such as the existence of numbers, properties, and the like. Instead of pondering whether numbers or properties (or mereological sums, or tables) exist, one should set up a linguistic framework, with explicit rules of grammar and rules of inference (and perhaps a model-theoretic consequence relation) for this purpose, and then see how useful the resulting framework is for various theoretical aims we may have.

However, scientists and, as noted above, mathematicians at work also do not begin their deliberations by first laying down rules for a linguistic framework, or a formal deductive system. Like the metaphysician, they use the informal languages they have inherited from their parents and teachers, and work in those. Is Carnap, or our Carnapian, saying that before anyone can ponder whether there are atoms, viruses, black holes, irrational numbers, infinitesimals, large cardinals, non-constructible sets, etc., she should first explicitly lay down a linguistic framework? Failing that, are we left with meaningless pseudo-questions?

We see two interpretive options, either for Carnap himself or for Carnapians, like ourselves. The first would put natural science and mathematics, as practiced, within the target, along with traditional metaphysics. Perhaps the Carnapian is proposing the linguistic frameworks as a sort of *explication*

---

[12]Note that Carnap here speaks of 'syntactical rules', while the present focus is on *semantics* – the meanings of logical terminology in particular. This does not represent any tension, however. Carnap's 'syntactical rules' are not mere statements of formation rules, or matters of grammar. Indeed, his main instances of syntactic rules are rules of inference. At least some logicians, the inferentialists, argue that inference rules – introduction and elimination rules in particular – give the meaning of logical terminology. Moreover, Carnap himself eventually allowed something like model-theoretic truth-conditions to figure in linguistic frameworks. See, for example, Carnap (1947). Thanks to a referee for highlighting this.

of actual deductive practice in mathematics, and actual deductive and empirical practice in the natural sciences.

The idea behind explication is that we ought to *replace* imprecise and informal concepts with rigorous ones (see Carnap 1950b). On this picture, the Carnapian is proposing that, instead of the loose and familiar, but informal standards of contemporary mathematical and scientific practice, the community *ought to* adopt an explicit linguistic framework, and be careful to abide by the rules sanctioned by that framework. Any existence questions, over atoms and infinitesimals for example, are answered internally, and trivially. The only important and legitimate external question is whether the various frameworks are useful to us (for whatever purposes we may have).

To be sure, it is not at all clear whether Carnap himself would conceive of linguistic frameworks as themselves explications. For one thing, he only provides a notion of explication for *concepts*. For Carnap, the 'task of explication consists in transforming a given more or less inexact concept into an exact one or, rather, in replacing the first by the second' (Carnap 1950b, 3). For example, one might explicate the concept of warmth with something like 'mean kinetic energy' or the more sophisticated scientific notions used today. Conceiving of a linguistic framework as an explication of a deductive practice, as this (rather tentative) proposal suggests, involves extending explication beyond concepts to something like ongoing practices. But we need not quibble over terminology here.

The key, on this first interpretive option, is that our Carnapian is making a *proposal to* mathematicians and scientists. Instead of working in their familiar, but informal language, they are admonished to explicitly adopt a linguistic framework, and use that instead. This would resolve the foregoing issues concerning the normativity of logic in a rather straightforward way, well within the purview of our pluralism. The claim is that the logic in question, be it classical, intuitionistic, or something else, is to be developed in rigorous linguistic framework, and that the practicing mathematician or scientist *ought to* adopt it, instead of her more loose and familiar forms of discourse and reasoning.

To put this in the terms of the previous section, this first option is, we think, too prescriptive, both as an interpretation of Carnap and also for present purposes (i.e. the truth of the matter). Carnap's first sketch of a linguistic framework is what he calls a 'thing language', adopted for discussing ordinary objects. He wrote that 'we all have accepted the thing language early in our lives as a matter of course' (Carnap 1950a, 243). But, whoever 'we' may be, we never did adopt a rigorous framework, with explicitly given rules, for talking about ordinary things. Or at least we did not do so 'early in our lives' (unless one's first logic course counts as 'early'). Also, Carnap's explicit target is traditional metaphysics, not science and mathematics. Indeed, traditional metaphysics is *contrasted* with practicing science.

The idea seems to be that the norms of rigor and deduction that underlie the actual development of mathematics and science are a lot closer to an explicitly defined linguistic framework than whatever norms and principles underlie traditional metaphysics (if not ordinary talk of ordinary things). In the terms of the previous section, one might think of a given linguistic framework as a *descriptive theory* of the norms of rigor that underlie a chunk of mathematics and perhaps science. They are out to codify the norms that are implicit in the practice itself. And the framework is accurate to the extent that it captures the relevant norms.

This steadfastly descriptive course may not be quite right either, although we submit that it is at least close to being right. The descriptive perspective here makes logic the same sort of enterprise as empirical semantics, hostage to the subtle details of how the logical terms are used in practice – in mathematics in the present case. There is often a gap between an explicit formal theory and the meanings of the logical terms actually used. Like most scientists, the logician idealizes.[13]

We need not take sides on whether a given logical system is an accurate description of all and only the norms that underlie a given branch of mathematics. Perhaps the best course, concerning the normativity of logic, is a sort of compromise between the prescriptive and the descriptive orientation toward linguistic frameworks – but we remain a lot closer to the descriptive side of the divide than the prescriptive. The compromise is to think of a linguistic framework as a rational reconstruction of the norms implicit in various branches of mathematics and perhaps science – and not as a proposed revision to those very norms. The logician will not tolerate large and significant gaps between logical theory and mathematical practice (Corcoran 1973).

Our modest proposal is thus like the one suggested by Burgess (1992, 2015), but without the limiting focus to classical mathematics. There are indeed norms implicit in the practices of classical analysis, intuitionistic analysis, inconsistent mathematics,... These are all evaluative practices, and the various logics are rational reconstructions of the norms implicit in each one, under the usual idealizations that go with empirical theorizing about evaluative practices.

The reason why there is no One True Logic is that the norms implicit in these practices are different from each other. Each of the logics is truth preserving in its domain, in the sense that if the premises of valid argument (in the indicated logic) are true in the domain, then so is its conclusion.

---

[13] See Glanzberg (2015) for an insightful account of the connection between logic and the meanings of natural language expressions.

None of the logics, or perhaps only the weakest of them, is truth preserving across the board.[14]

Acknowledgements

We would like to thank Robert Kraut, Kevin Scharp, audiences at the Society for Exact Philosophy meeting in 2017, a special session at the Pacific APA organized by the Relativisms Global Research Network in 2017, the graduate student workshop at Ohio State, and two referees for helpful feedback.

Disclosure statement

No potential conflict of interest was reported by the authors.

References

Anderson, A. R., and N. D. Belnap. 1975. *Entailment: The Logic of Relevance and Necessity*, Vol. I. Princeton, NJ: Princeton University Press.
Beall, J., and G. Restall. 2006. *Logical Pluralism*. Oxford: Clarendon Press.
Benacerraf, P., and H. Putnam, eds. 1983. *Philosophy of Mathematics*. Cambridge: Cambridge University Press.
Bishop, E. 1967. *Foundations of Constructive Analysis*. New York: McGraw-Hill.
Bishop, E., and D. Bridges. 2012. *Constructive Analysis*. Grundlehren der mathematischen Wissenschaften. Berlin: Springer.
Bridges, D. 1979. *Constructive Functional Analysis*. Research Notes in Mathematics. London: Pitman.
Brouwer, L. 1913. "Intuitionism and Formalism." *Bulletin of the American Mathematical Society* 20 (2). Reprinted in Benacerraf and Putnam(1983), 77–89.
Burgess, J. 1992. "Proofs about Proofs: A Defense of Classical Logic." In *Proofs, Logic and Formalization*, edited by M. Detlefsen, 8–23. London: Routledge.
Burgess, J. 2015. *Rigor and Structure*. Oxford: Oxford University Press.

---

[14] One might wonder if there a connection between the foregoing pluralism (or pluralisms) about logic and a pluralism about truth. Is, say, truth in classical arithmetic the same thing, or the same kind of thing, as truth in intuitionistic analysis or smooth infinitesimal analysis? We do not have strong views on this, or on the nature of truth generally. There are some programs of truth-pluralism, due to thinkers like Wright (1992) and Lynch (2009). The idea, roughly, is that there are different properties that manifest truth in different discourses. Typically, advocates of views like this say that mathematics, as a whole, is *one* such discourse, in which case there would be only one property that manifests truth for mathematics.
One can get started on a kind of truth pluralism, to go with the foregoing logical pluralism, by modifying that, and allowing that there are different mathematical discourses, each with its own truth property. For example, the truth property for classical mathematics might be a kind of correspondence (as is typically found in science and ordinary discourse about ordinary things, according to most truth pluralists). The truth property for constructive and intuitionistic mathematics might be an anti-realist one, such as super-assertibility.
The problem here is that these are not the only two options. Smooth infinitesimal analysis invokes intuitionistic logic, but it does not seem to be motivated by anti-realism (see, e.g. Hellman 2006; Shapiro 2014; Rumfitt 2015, §7.4). The truth pluralist would have to identify a property that manifests truth in this branch of mathematics. And we have no clue as to what the truth property would be for inconsistent mathematics, or in any other discipline. We hope to pursue this matter in future work.

Carnap, R. 1937. *The Logical Syntax of Language*. New York: Harcourt, Brace and Company.
Carnap, R. 1947. *Meaning and Necessity*. Chicago, IL: University of Chicago Press.
Carnap, R. 1950a. "Empiricism, Semantics and Ontology." *Revue Internationale de Philosophie* 4 (11): 20–40.
Carnap, R. 1950b. "On Explication." In *Logical Foundations of Probability*, 1–18. Chicago, IL: University of Chicago Press.
Coffa, J. 1993. *The Semantic Tradition from Kant to Carnap*. Cambridge: Cambridge University Press.
Corcoran, J. 1973. "Gaps between Logical Theory and Mathematical Practice." In *The Methodological Unity of Science*, edited by M. A. Bunge, 23–50. Boston: Reidel.
Dummett, M. 1973. "The Philosophical Basis of Intuitionistic Logic." In *Truth and Other Enigmas*, edited by M. Dummett, 215–247. Duckworth.
Field, H. 2006. "Truth and the Unprovability of Consistency." *Mind* 115 (459): 567–606.
Frege, G. 1893. *Basic Laws of Arithmetic*. Oxford: OUP. Translated as *Basic Laws of Arithmetic*, by Philip A. Ebert and Marcus Rossberg, with Crispin Wright. Oxford: Oxford University Press; 2013.
Glanzberg, M. 2015. "Logical Consequence and Natural Language." In *Foundations of Logical Consequence*, edited by C. Caret and O. Hjortland, 71–120. Oxford: Oxford University Press.
Hellman, G. 2006. "Mathematical Pluralism: The Case of Smooth Infinitesimal Analysis." *Journal of Philosophical Logic* 35 (6): 621–651.
Heyting, A. 1956. *Intuitionism*. Amsterdam: North-Holland.
Keefe, R. 2014. "What Logical Pluralism Cannot Be." *Synthese* 191: 1375–1390. Kouri, T. 2016. "Logical Instrumentalism." PhD thesis, Ohio State University.
Lynch, M. P. 2009. *Truth as One and Many*. Oxford: Oxford University Press.
MacFarlane, J. 2002. "Frege, Kant, and the Logic in Logicism." *Philosophical Review* 111 (1): 25–65.
MacFarlane, J. 2004. "In What Sense (If Any) is Logic Normative for Thought?" Unpublished.
Prior, A. 1967. "The Runabout Inference Ticket." In *Analysis*, edited by P. Strawson, 38–39. Oxford: Oxford University Press.
Read, S. 2006. "Review of Logical Pluralism." In *Notre Dame Philosophical Reviews*. http://ndpr.nd.edu/news/logical-pluralism/.
Richman, F. 1996. "Interview with a Constructive Mathematician." *Modern Logic* 6(3): 247–271.
Rumfitt, I. 2015. *The Boundary Stones of Thought: An Essay in the Philosophy of Logic*. Oxford: Oxford University Press.
Searle, J. 1969. *Speech Acts: An Essay in the Philosophy of Language*. Cam: Verschiedene Aufl. Cambridge University Press.
Searle, J. 2010. *Making the Social World: The Structure of Human Civilization*. Oxford: Oxford University Press.
Shapiro, S. 2014. *Varieties of Logic*. Oxford: Oxford University Press.
Steinberger, F. 2017. "Frege and Carnap on the normativity of logic." *Synthese* 94: 143–162.
Tennant, N. 1996. "The Law of Excluded Middle is Synthetic A Priori, If Valid." *Philosophical Topics* 24 (1): 205–229.
Williamson, T. 2013. *Modal Logic as Metaphysics*. Oxford: Oxford University Press.
Wright, C. 1992. *Truth and Objectivity*. Cambridge, MA: Harvard University Press.

Disagreement about logic

Ole Thomassen Hjortland

ABSTRACT
What do we disagree about when we disagree about logic? On the face of it, classical and nonclassical logicians disagree about the laws of logic and the nature of logical properties. Yet, sometimes the parties are accused of talking past each other. The worry is that if the parties to the dispute do not mean the same thing with 'if', 'or', and 'not', they fail to have genuine disagreement about the laws in question. After the work of Quine, this objection against genuine disagreement about logic has been called the meaning-variance thesis. We argue that the meaning-variance thesis can be endorsed without blocking genuine disagreement. In fact, even the type of revisionism and nonapriorism championed by Quine turns out to be compatible with meaning-variance.

1. The meaning-variance thesis

Suppose that two logicians, Astrid and Beatrice, are arguing about a logical law. Astrid says that the law is valid, Beatrice that it is invalid. Both of them have reasons for their beliefs. They believe what they believe because they subscribe to different theories about logical properties such as validity, truth, and provability. At least on the face of it Astrid and Beatrice are capable of having a rational debate about their beliefs. They offer arguments, objections, and replies. Although they ultimately stick to their respective positions, they feel the pull of each other's arguments. Every now and then Astrid relies on the very law in dispute when arguing for her position. Beatrice rejects these arguments. Nonetheless, they don't give up on finding ways of convincing the other person that they in fact have the superior position. Despite the apparent stalemate, Astrid and Beatrice both think the debate is worthwhile.

The situation, I believe, is familiar in the philosophy of logic. Classical and nonclassical logicians disagree about logical laws, and they offer reasons for their positions. Sometimes the arguments rely on contentious principles, but not always. The participants, myself included, believe that the debate is important and productive. Nonetheless there is a well-known claim to the effect that there is no genuine disagreement in cases such as that of Astrid and Beatrice. Instead they are involved in a mere verbal dispute. Classical and nonclassical logicians are not engaged in a substantive debate about the nature of logical laws, but are simply attaching different meanings to the same expressions. Once the parties are clear on what they mean by locutions such as 'and', 'not', 'valid', 'proof', the conversation can proceed with the dispute resolved.

This claim has been called the *meaning-variance thesis*, and it is attributed to Quine (1960a, 1986).[1] In a pivotal section, entitled 'Change of logic, change of subject', Quine considers a 'popular extravaganza', a discussion between a logician who rejects the law of non-contradiction and a counterpart who warns against the consequences:

> My view of the dialogue is that neither party knows what he is talking about. They think that they are talking about negation, '$\sim$', 'not'; but surely the notion ceased to be recognisable as negation when they took to regarding some conjunctions of the form '$p.\sim p$' as true, and stopped regarding such sentences as implying all others. Here, evidently, is the deviant logician's predicament: when he tries to deny the doctrine he only changes the subject. (Quine 1986, 81)

Elsewhere Quine puts the point in a different way:

> [T]he departure from the law of the excluded middle would count as evidence of revised usage of 'or' and 'not' .... For the deviating logician the words 'or' and 'not' are unfamiliar, or defamiliarised. (Quine 1960a, 362)

So when the nonclassical logician rejects a central classical law, say, for negation, the use of the expression has been revised. '[A]lternative logics are inseparable practically from mere change in usage of logical words.' (Quine 1960a, 355)

In one sense it is trivial that rejecting a classical logical law will lead the nonclassical logician to revise the use of an expression. If one rejects the law of excluded middle, then the use of disjunction should reflect that. But Quine's claim appears to be stronger. He says that neither party knows what they are talking about. The nonclassical logician 'only changes the subject'. Nonclassical logic cannot be separated from 'mere change in usage'.

---

[1] As far as I know the name is due to Haack (1978, 224).

Quine (1986, 81) gives an example to motivate the thesis. He asks us to consider a deviant logician who has simply swapped around the logical laws of conjunction and disjunction. In that case, he suggests, it would be 'nonsense' to think that the deviant logician really disagreed with us 'on points of logical doctrine'. He continues:

> There is no residual essence of conjunction and alternation in addition to the sounds and notations and the laws in conformity with which a man uses those sounds and notations. (Quine 1986)

Although Quine doesn't talk about change of meaning explicitly, others have interpreted him as saying that a change of logical laws involves a change of meaning. The thinking is that the laws that govern a logical expression –or at least some of them – also fix the meaning of the expression. Thus, if a nonclassical logician rejects modus ponens for the conditional, the resulting nonclassical conditional differs in meaning from the conditional of classical logic.

Haack (1974, 8) explicitly attributes an argument of this form to Quine. Before quoting Quine as an example, she points out that some authors have claimed that classical and nonclassical logics are not genuine rivals, 'because their apparent incompatibility with classical logic is explicable as resulting from *change of meaning* of the logical constants'. Dummett (1978a) also construes Quine's argument as an argument about meaning:

> [Quine] begins by remarking that it is impossible for anyone to deny a law of classical logic: for, if he fails to accept some formula which a classical logician would take to be a formulation of such a law, this failure would establish a conclusive ground for saying that he was not attaching to the logical constants appearing in the formula the same meanings as those attached by the classical logician, and hence he had not denied anything held by the classical logician, but merely changed the subject. (Dummett 1978a, 270)

The worry is that there is no proposition that the classical and nonclassical logician disagree about. Despite the apparent disagreement about, say, whether the law of excluded middle is valid, the parties are expressing their opinion about different propositions. The classical logician attributes classical validity to a proposition formed with their negation and disjunction, while the nonclassical logician attributes nonclassical invalidity to a proposition formed with their negation and disjunction. The result, Haack warns, is that 'there is no real conflict between [nonclassical] and classical logic' (Haack 1974, 8).

2. Quine's revisionism and gradualism

Both Haack and Dummett argue that Quine's meaning-variance thesis is at odds with the view about logic he championed in his influential 'Two Dogmas of Empiricism' (Quine 1951). Here Quine claims that no theoretical statement is immune to revision, not even the laws of logic. This is Quine's *revisionism*:

> Conversely, by the same token, no statement is immune to revision. Revision even of the logical law of the excluded middle has been proposed as a means of simplifying quantum mechanics; and what difference is there in principle between such a shift and the shift whereby Kepler superseded Ptolemy, or Einstein Newton, or Darwin Aristotle? (Quine 1951, 40)

Quine's question is meant to suggest that there is only a difference in degree between an allegedly *a priori* theory such as logic and the theories of the natural sciences. This is Quine's *gradualism*:

> The kinship I speak for is rather a kinship with the most general and systematic aspects of natural science, farthest from observation. Mathematics and logic are supported by observation only in the indirect way that those aspects of natural science are supported by observation; namely, as participating in an organised whole which, way up at its empirical edges, squares with observation. I am concerned to urge the empirical character of logic and mathematics no more than the unempirical character of theoretical physics; it is rather their kinship that I am urging, and a doctrine of gradualism. (Quine 1986, 100)

This kinship between logic and the empirical sciences have been echoed by a number of more recent authors, for example Maddy (2002, 2014), Priest (2006, 2014, 2016), Russell (2014, 2015, 2018), Williamson (2017), and Hjortland (2017a, 2017b). Although they differ on the details, all these authors reject the privileged epistemological status of logic, be it foundationalism, apriorism, obviousness or self-evidence. Instead they subscribe to a view where logical theories are supported by arguments similar to those used for theories in other sciences. That doesn't mean that logic is directly empirical in the sense that its laws are justified by induction, but it does mean that logical theories answer to *a posteriori* evidence through their connections with empirical theories.

According to revisionism, there can be genuine disagreement about logical theories, and furthermore, there can be arguments that support rational theory-choice. So if Quine favours a form of anti-exceptionalism, how can that be squared with his meaning-variance thesis? Dummett's answer is that Quine has simply abandoned his position from 'Two Dogmas' in his later writings (e.g. Quine 1960b, 1986). 'In the meantime',

Dummett writes, 'Quine himself has totally reversed his position, as may be seen from the chapter on 'Variant Logics' in *Philosophy of Logic*' (Dummett 1978b, 270). Haack also concludes that Quine has abandoned his earlier position:

> The attack in 'Two Dogmas' on synonymy etc. *would* threaten an account of logical truths as analytic because *true in virtue of the meaning of the logical constants*. Now in *Word and Object* Quine renews his sceptical attack on meaning notions, but makes an exception in the case of the logical connectives, which, he claims, do have determinate meaning ... ; and this paves the way for his acceptance ... of a meaning-variance argument to the effect that the theorems of deviant and classical logics are, alike, true in virtue of the meaning of the (deviant or classical) connectives. (Haack 1978, 237)

If Haack and Dummett's interpretation is correct, Quine simply abandoned a key component of his revisionism when he gave his meaning-variance thesis.[2] If Dummett is right, and it is '*impossible* for anyone to deny a law of classical logic' (Dummett 1978b, 270, my emphasis), then how can one simultaneously insist that logic is not only revisable, but rationally revisable?

Hence, revisionists have a reason to reject the meaning-variance argument. Indeed, many commentators have found fault with Quine's argument for the thesis, regardless of whether or not they agree on Quine's particular gradualist picture.[3] In the next section, we will look at some prominent objections to the meaning-variance thesis.

3. Against meaning-variance

Recall Astrid and Beatrice, the two logicians who are in a dispute about a logical law. Suppose that their dispute is about modus ponens (in the form $A, A \to B \vDash B$). Astrid claims that it is valid, Beatrice disputes it. Suppose further that modus ponens is a meaning-determining principle for Astrid's conditional. Since modus ponens is meaning-determining, it is valid simply in virtue of the meaning of Astrid's conditional. The worry is that Beatrice, in disputing modus ponens, doesn't succeed in rejecting the law that Astrid supports. For that law is expressed with Astrid's conditional, and if Beatrice disputes it, she is simply rejecting some other

---

[2]Other authors have also attributed two mutually exclusive positions to Quine. Williamson (2007, 97) says that 'Quine's epistemological holism in 'Two Dogmas' undermines his notorious later claim about the deviant logician's predicament'. Arnold and Shapiro (2007) investigate the conflict between what they call the 'logic-friendly Quine' and the 'radical Quine'.
[3]Of course, Dummett himself is a revisionist about logic. More about that below.

law, expressed with her own Beatrice-conditional (for which modus ponens doesn't hold). There is, therefore, no single proposition – *the logical law* – that Astrid accepts and Beatrice rejects.

If it is indeed true that Beatrice does not succeed in disputing the law that Beatrice accepts, and, moreover, genuine disagreement is propositional disagreement, then there is no genuine disagreement between Astrid and Beatrice. The same line of reasoning will carry over to disputes about any logical law that is meaning-determining. This is what leads Haack (1974, 8) to conclude that there is 'no real conflict' between the parties of the dispute. She outlines the argument as follows:

(a) if there is change of meaning of the logical constants, there is no real conflict between Deviant and classical logic,
(b) if there is Deviance, there is change of meaning of the logical constants, so
(c) there is no real conflict between Deviant and classical logic. (Haack 1974)

However, it is unclear what Haack means by 'no real conflict'. So in what follows, we will work with a reformulation of the argument better suited to our purposes:

(P1) If two parties, A and B, are in a dispute over a basic logical law S involving logical expressions $o_1, \ldots, o_n$, then their dispute arises wholly in virtue of A and B's disagreement about the meaning of $o_1, \ldots, o_n$.
(P2) If A and B are in a dispute over S, and the dispute arises wholly in virtue of their disagreement about the meaning of an expression o in S, then their dispute is verbal.
(C) If two parties, A and B, are in a dispute over a basic logical law S involving logical expressions $o_1, \ldots, o_n$, then their dispute is verbal.

The first premise (P1) corresponds very roughly to Haack's premise (b), but the formulation is different in a couple of ways. First, P1 talks only about disputes over *basic logical laws*. The point is that two logicians could be in a dispute about whether or not something is a logical law (a theorem), agree on all axioms and primitive proof rules, but nonetheless reach different conclusions because one of them made an error. In such cases, one would hope that the error could be spotted, leading to genuine agreement. Those are clearly not the sort of cases Quine's meaning-variance thesis is supposed to apply to. Thus, P1 limits the disputes to those about a subclass of logical laws.

Furthermore, note that basic logical laws can refer to logical truths (e.g. the law of non-contradiction $\models \neg(\varphi \wedge \neg\varphi)$) and to valid arguments (e.g. modus ponens in the form $\varphi, \varphi \to \psi \vDash \psi$), but also to valid meta-arguments, that is, arguments from a set of arguments to an argument (e.g. conditional proof, classical *reductio*).[4] This is important because many of the actual disputes between classical and nonclassical logics are about meta-arguments. In fact, certain logics agree on the validity of all arguments, but disagree about meta-arguments.

Second, in Haack's (b) we only learn that logical 'deviance' leads to 'change in meaning'. This is misleading. As Priest (2006) points out, even if we accept the difference in meaning, it doesn't follow that the classical logician keeps the meaning fixed, whereas the nonclassical (i.e. deviant) logician 'changes' the meaning. The classical logician certainly is not entitled to claim that their logical concepts are the standard logical concepts, for example in the sense that natural language expressions pick out the classical concepts. That is precisely one of the claims that has been disputed by nonclassical logicians. Rather, the two parties of the dispute enter the dispute with different meanings attached to the same logical expression.

Third, P1 says that the difference in meaning is due to the expressions involved in the logical law. For example, if the law of excluded middle ($\varphi \vee \neg\varphi$) is basic, a dispute could arise in virtue of a disagreement about $\vee$ or $\neg$. However, the dispute could also arise from a disagreement about the meaning of meta-theoretic expressions. The law that the dispute is about is better expressed as:

(1) $\ulcorner \varphi \vee \neg\varphi \urcorner$ is valid.

In that case, the two parties to the dispute could be assigning the same meaning to $\vee$ and $\neg$, but different meanings to the logical expression 'valid'.[5] After all, classical and nonclassical logics mean different things by these expressions, just like they do with the run of the mill connectives. They might agree on theory-relative terms, such as 'classically valid' or

---

[4]Conditional proof and classical *reductio* respectively:

$$\dfrac{\begin{array}{c}[\varphi]^u\\ \vdots\\ \psi\end{array}}{\varphi \to \psi}(u) \qquad \dfrac{\begin{array}{c}[\neg\varphi]^u\\ \vdots\\ \bot\end{array}}{\varphi}(u)$$

[5]A simple example is a pair of many-valued systems with the same truth-functions attached to the logical connectives, but where validity is defined in terms of different preservation properties. See Hjortland (2012) for a more detailed example.

'intuitionistically provable', but they do not seem to agree on the meaning of 'valid' and 'provable' simpliciter.[6] The result is that Quine's meaning-variance thesis arguably applies not only to logical connectives and quantifiers, but also to meta-theoretic logical expressions (e.g. 'valid', 'consistent', 'provable'). It is convenient, therefore, to talk more broadly about logical laws as including such meta-theoretical claims, even if that is unorthodox terminology.

Just as P1 corresponds to Haack's premise (b), P2 roughly corresponds to Haack's (a). However, P2 avoids saying that there is 'no real conflict' between classical and nonclassical logics, and instead claims that disputes about basic logical laws are *verbal disputes*. In fact, P2 is a slightly modified version of an account of verbal disputes due to Chalmers (2011). This account does not say anything about logical disputes in particular, but gives a sufficient condition for a dispute in general to be verbal. For the purposes of the meaning-variance thesis, we will simply assume that Chalmers' account is correct. What counts as a verbal dispute is certainly an interesting discussion, but it is beyond the scope of this paper.

For our purposes it is therefore the first premise that will receive most of the attention. If P1 turns out to be false, then at least this type of argument for the meaning-variance thesis is blocked. This is indeed the strategy that a number of authors have used in their objection to Quine's claim. What is required, then, is a counterexample to the the conditional of P1: a case where two parties *A* and *B* are in dispute over a basic logical law *S*, but where their dispute doesn't arise wholly in virtue of *A* and *B*'s disagreement about the meaning of the logical expressions involved.

Suppose that modus ponens is a basic logical laws for the conditional, and that *A* and *B* are in dispute over modus ponens. First, note that if *B* rejects modus ponens, it is compatible with P1 that *B believes* that the dispute does not arise wholly in virtue of a disagreement about meaning. She might insist that the meaning she assigns to the conditional is the same as *A*, despite the fact that she rejects modus ponens. All that P1 requires is that the dispute is, in fact, a result of a disagreement about meaning, even if *B* fails to realise it. She might have been convinced, for example, by McGee's (1985) counterexample to modus ponens, but still insist that she shares the same meanings as *A*. Of course, it could turn out that what McGee's counterexample shows is that, contrary to previous claims, modus ponens isn't a basic logical law for the conditional; indeed, that it isn't a logical law at all! But that is besides the point here. An

---

[6]See Field (2015) for a discussion about this distinction.

objection to P1 would have to satisfy the antecedent, namely that S is a basic logical law.

One way to achieve this is to divorce basic logical laws from meaning-determining laws. The law in dispute is only said to be a basic law, so if modus ponens is a basic logical law for the conditional (in the sense that it is primitive), but without being meaning-determining, then rejecting it is compatible with full agreement on meaning. In other words, P1 can be resisted.

An early proponent of this objection is Putnam (1976). In keeping with Quine's revisionism, Putnam suggests that quantum mechanics gives us reason to revise classical logic in favour quantum logic.[7] In the latter system, the law of distributivity (i.e. $(\varphi \wedge (\psi \vee \chi)) \rightarrow ((\varphi \wedge \psi) \vee (\varphi \wedge \chi))$) is rejected, and so the question of whether we are changing the meaning of disjunction arises. However, Putnam believes that the revision is compatible with sameness of meaning. He observes that even if there is subtle change of the primitive proof rules for disjunction, there is a substantial number of shared theorems between classical and quantum logic. For Putnam, the problem is that 'we do not posses a notion of 'change in the meaning' refined enough to handle this issue' (Putnam 1976, 190):

> Only if it can be made out that [distributivity] is 'part of the meaning' of 'or' and/or 'and' (which? and how does one decide?) can it be maintained that quantum mechanics involves a 'change in the meaning' of one or both of these connectives. (Putnam 1976)

In a similar vein, Field (2008, 17) also rejects the meaning-variance thesis because of doubts about the very distinction between meaning-determining and non-meaning-determining laws:

> The question [of meaning change] is clear only to the extent that we know how to divide up such firmly held principles into those that are 'meaning constitutive' or 'analytic' and those which aren't, and this is notoriously difficult.

The worry is echoed by Shapiro (2014, 91):

> My tentative conclusion ... is that the talk of whether the meaning of the logical terminology is the same or different in different contexts/theories is itself context-sensitive and, moreover, interest-relative – in part because such talk is vague. So I am sceptical of the very question of whether the classicist and intuitionist, for example, talk past each other – if we insist that this is a sharp and objective matter, to be determined once and for all.

---

[7] Putnam later changes his mind about quantum logic (cf. Putnam 1981, 2012).

A related criticism of meaning-variance is that although there are meaning-determining basic logical laws, not all basic laws are meaning-determining. Elsewhere I have called this view *minimalism* about the meaning of logical connectives (Hjortland 2012, 2014). Roughly put, the minimalist acknowledges that a proper subset of the logical laws are meaning-determining, but allows for non-verbal disputes about the remaining laws. The minimalist view has an early formulation in the work of Putnam (1957) and Haack (1974), but has been given a more precise formulation by Paoli (2003, 2007, 2014) and Restall (2002). Here is Restall (2002, 11):

> If any set of rules is sufficient to pick out a single meaning for the connective, take that set of rules and accept those as meaning determining. The other rules are important when it comes to giving an account of a kind of logical consequence, but they are not used to determine meaning.

Again, such a division of labour between logical laws will allow for disputes that are non-verbal, though only if the disputes are about non-meaning-determining laws. The difficulty is to identify a non-arbitrary divide between the basic laws that contribute to the meaning and the ones that do not.

Paoli has developed an answer to this question by applying a useful distinction from structural proof theory. In sequent calculus rules there are two types of proof rules: *operational rules*, that is, left and right proof rules that govern specific logical expressions, and *structural rules*, that is, proof rules that govern the deducibility relation independent of the type of formula in question (e.g. weakening, contraction).[8] Paoli's proposal is that whereas operational rules are meaning-determining, structural rules are not. With the distinction in place, the meaning theory allows for non-verbal disputes between logical theories that differ only with respect to structural properties. That includes linear logics, affine logics, and a number of relevant logics.

An immediate worry is that some standard disputes, such as that between classical and intuitionistic logic, won't count as genuine disagreements on this account. But, since classical and intuitionistic logic can be formalised in sequent calculus in such a way that the only difference is the cardinality of the succedent, it has been suggested that even this dispute would count as a genuine disagreement under the proposal.[9] However, such a fix raises a more general concern. If we anchor the

---

[8]See Negri and vonPlato (2001) for an introduction.
[9]The point is anticipated by Haack (1974).

meaning-determining versus non-meaning-determining distinction to the operational versus structural distinction, we need the latter distinction to be sufficiently stable. The problem is that what counts as structural is highly system-dependent. Different sequent calculi introduce structural operators in the proof theory in order to add structural rules. In standard sequent calculus we have right- and left-side commas, while generalisations like hypersequents or labelled sequent calculus allow for new structural rules. Bunch theory with intensional and extensional premise-combinators and bilateral systems further load the structural language with operators. Such systems simply shift the various logical properties of connectives over to the structural language.

In addition, even if the structural rules can be claimed to be independent of the meaning of the specific connectives, there is still the question of whether or not a change in structural rules brings about a shift in the meaning in metatheoretic expressions such as 'valid' or 'provable'. If that is the case, and the disputes are about the validity of, say, the law of excluded middle, we are no closer to an account of nonverbal disputes.[10]

Let us return to the argument for the meaning-variance thesis. Are there other strategies for resisting the premise P1? Another influential criticism of the idea that there are meaning-constitutive principles is due to Williamson (2007). He formulates a number of objections against so-called understanding-assent links (UA links). He denies that there are logical laws that a speaker must assent to in order to qualify as understanding a connective. More generally, let S be a sentence that expresses such a principle. Then consider the following UA link:

(2) Necessarily, everyone who understands the sentence S assents to it.

Contraposed, everyone who dissents to S fails to understand it. Williamson objects to (2) and a number of related UA links. To see one reason why consider the following example. Suppose that modus ponens is the principle in question. Recall that McGee (1985) thinks there are counterexamples to modus ponens, and therefore rejects the validity of the law. By the UA link this entails that McGee does not understand the English language expression 'if'. Williamson thinks this conclusion is simply false:

> In conversation, [McGee] appears to understand it perfectly well. By ordinary standards, he *does* understand it. Before he had theoretical doubts about

---

[10]See Hjortland (2014) for a more detailed criticism of minimalism, and Paoli (2014) for some promising replies on behalf of the minimalist.

modus ponens, he understood it. Surely his theoretical doubts did not make him cease to remember what it means. Moreover, his doubts derive from taking at face value a natural pattern of native speaker reactions to an ingeniously chosen case. If he counts as not understanding 'if ', so do millions of other native speakers of English. (McGee 1985, 94)

If it turns out that the premise P1 entails the UA link, Williamson's argument would offer a strategy for blocking the argument for the meaning-variance thesis. If we have reason to reject the UA link, we would also have reason to reject P1. However, in the current formulation P1 does not entail the UA link. In fact, while it does require that one party dissents to a logical principle, it does not require that the party therefore fails to understand any of the involved logical expressions. It simply says that the dispute arises wholly in virtue of their disagreement about the meaning of the expressions.

Of course, one could object that what Quine *really* had in mind was something stronger than P1, and indeed a premise that entails the UA link. That is not entirely uncalled for given Quine's verdict that 'neither party knows what he is talking about' in disputes about logical principles. At other points, however, Quine makes comments that point to the more modest formulation. It is a formulation, I argue, that allows us accept the meaning-variance thesis without jettisoning genuine disagreement and the revisionism he endorses in his earlier work.

4. Revisionism and meaning-variance reconciled

Not everyone has concluded that there is a dissonance between Quine's revisionism and the meaning-variance thesis. Although Priest (2006) disagrees with Quine's classicism, he suggests that there is more of a continuity between Quine's views than he is given credit for:

> It would be wrong, however, to suppose that Quine takes the meaning variance of connectives across different logics to be an argument against his earlier view concerning the revisability of logic. (Priest 2006, 168)

Priest points out that rather than rejecting his old position, Quine (1986) seems to open for the possibility that meaning-variance is not at odds with rational revision of logical laws. Only a few pages after his infamous 'change of logic, change of subject' quip, Quine is again discussing revision of a classical law, but now in a more concessive manner:

> [W]hoever denies the law of excluded middle changes the subject. This is not to say that he is wrong in so doing. In repudiating 'p or $\sim p$' he is indeed giving up

classical negation, or perhaps alternation, or both; *and he may have his reasons.* (Quine 1986, 83, my emphasis)

The contrast with the more polemical comments from earlier in the same chapter is striking. Rather than accusing the parties of speaking past each other, he now suggests that rejecting classical laws could have a rationale, even if it involves giving up classical logical concepts.

But if revisionism is possible despite meaning-variance, what is the nature of the revision and what is the nature of the underlying disagreement? One possibility is that the disagreement is simply *about* meaning. By raising the disagreement to a metalinguistic level, we can own the meaning-variance thesis and still save genuine disagreement. Granted, in that case the impression that the disagreement was *about* the law of excluded middle is misleading.

That logical disputes are sometimes about meaning should be no surprise. Some standard cases of disputes in the philosophy of logic are overtly about meaning. Some discussions about logical laws are about how natural language expressions, for instance 'not', 'if', 'all', are actually used. Semantic projects in the spirit of Montague (1973) and Davidson (1967) are typically attempts at using formal systems as tools in natural language semantics.[11] Similarly, critics of classical logic have pointed out that the logical laws governing, say, material implication or the boolean negation simply do not accurately reflect how we use their natural language counterparts. When relevantists refer to the postive and negative paradoxes, and when McGee gives a natural language counterexample to modus ponens, they are relying on natural language data. So, if the project is to give a formal semantics for certain natural language expressions (i.e. the logical expressions), the dispute is clearly about the meaning of these expressions.

But although formal logics are routinely used for this purpose, not all logical disputes are about the semantics of natural language. After all, classical and many nonclassical logics are not really concerned with the complexities of natural language usage. Instead, some logicians are more concerned with the logic of the concepts we *ought* to use than the logic of the concepts we actually use. The approach varies from those who merely suggest that scientific purposes call for a light modification of natural language (e.g. Quine's regimentation) to those who insist that a formal language should at least partly replace natural language in order

---

[11] Arguably the logic-as-modelling approach in Cook (2002) and Shapiro (1998, 2006) is a variant of the semantic approach, but one on which the logical systems allow for more idealisation.

to achieve sufficient precision (e.g. Frege's *Begriffschrift*). Shapiro (2006, 48) calls this the *normative orientation* towards logic.

A well-known example of the normative orientation is the type of intuitionist revisionism promoted by Dummett (1991). According to Dummett, the classical negation concept semantically misfires, and should therefore be replaced by intuitionistic negation. The argument is explicitly about meaning. In fact, Dummett's contention is precisely that a revisionist argument for logic must proceed via meaning-theoretic considerations. He identifies conditions that the inference rules for logical expressions must satisfy in order to count as meaningful (e.g. conservativeness, harmony), and argues that classical negation fails the test. Both the conditions that Dummett introduce and the claim that they aren't satisfied by classical concepts have been contended in the literature, but the details are not important here.[12] The point is that Dummett's revisionism is not only compatible with the meaning-variance thesis, but premised on the very idea that revision of logic springs out of a disagreement about which concepts are legitimate. The disagreement is metalinguistic, but definitely not 'merely verbal'. Compare it, for example, with Quine's case of merely verbal dispute where someone has simply swapped the terms '∨' and '∧' in the logical laws. The latter dispute results from a notational variant, while Dummett's revisions are motivated by meaning-theoretic arguments. Even if we ultimately reject his arguments for intuitionism, the revisionary strategy cannot be written off as a linguistic confusion.

5. Revision of concepts

Dummett's revisionism is actually an example of a philosophical methodology that has received more attention recently. Instead of merely offering conceptual analysis of crucial concepts such as personal identity, justice, truth, and knowledge, some philosophers engage in conceptual *engineering*.[13] They do not merely seek to elucidate the concepts we actually possess, but to investigate which concepts we *ought* to use for certain purposes. When concepts do their job well, they help us represent the world and express our thoughts. But concepts can also be unhelpful. They can be too imprecise or nonexpressive. By some accounts they can also be inconsistent or unethical. Since concepts can be deficient in a variety of ways,

---

[12]See for example Read (2000).
[13]See Scharp (2013), Burgess and Plunkett (2013), Cappelen (2018), and Eklund (2017a) for some recent approaches to conceptual engineering as a methodology. Tanswell (2018) has also explored how conceptual engineering can be applied in set theory. The label 'conceptual engineering' is due to Blackburn (1999).

the argument goes, we are sometimes better off with new concepts. Ultimately, those who promote conceptual engineering think that we sometimes ought to introduce replacement concepts.

Here are some examples. The folk concept *mass* is muddled. It runs together two theoretical concepts, *weight* (measured in newtons) and *mass* (measured in kilograms). The folk concept *fish* is unsuited for the theoretical purposes of biology, where it has been replaced by the typological concept *pisces* and then by concepts suited for phylogenetic classification. When greater precision is called for, the everyday concepts of *likelihood* and *chance* are dispensed with in favour of the mathematically precise concept of *probability*.

These are cases where deficient concepts have supposedly been replaced by more sophisticated, theoretical alternatives. In other cases conceptual engineering doesn't involve conceptual re-engineering of already existing concepts.[14] . Instead concepts that bear little similarity with already applied concepts are introduced to fulfil some purpose. Concepts such as *quark* in physics or *DNA* in biology do not serve as improvements of deficient concepts, but allow for the formulation of improved theories.

What about concepts in logical theories? When Dummett (1973, 454) talks about deficient concepts, he mentions the expression 'Boche', a slur used by the French against Germans. The example is supposed to anticipate his analysis of allegedly deficient logical concept, first the trivialising connective 'tonk' (Prior 1960), and ultimately the problems with classical negation. For Dummett, 'rejection or revision of concepts' is called for when the the inference rules associated with the concept fail to ' harmonise'. Although his diagnosis of the defects of logical concepts is idiosyncratic, it is natural to think of this and related projects in the philosophy of logic as cases of conceptual engineering.[15]

Later philosophers of logic have been even more explicitly engaged in what they consider conceptual engineering. The most widely discussed case is likely the concept of truth. A number of authors have defended *inconsistency theories*, according to which the everyday concept of truth is inconsistent (e.g. Azzouni 2007; Burgess and Burgess 2011; Eklund 2002, 2007; Patterson 2007, 2009; Scharp 2007, 2013; Scharp and Shapiro 2017). Although the authors disagree about whether or not the

---

[14]According to Brun (2016), conceptual re-engineering is closely related to Carnap's notion of explication (see also Novaes and Reck 2017).
[15]Tennant (1987, 1997) and Brandom (1994) are two philosophers who have followed Dummett in relevant respects.

alleged inconsistency is a reason to replace the everyday concept of truth, at least some inconsistency theorists have developed and defended replacement concepts. These concepts are designed to preserve consistency in classical logic, while simultaneously staying faithful to a number of the expressive roles traditionally assigned to the truth predicate.[16]

In order to be able to count concepts, and not only theories as inconsistent, inconsistency theories agree that concepts are governed by constitutive principles. A familiar example from logic is that standard proof rules are constitutive principles of logical concepts: conditional proof and modus ponens are constitutive of the conditional; disjunction introduction and proof by cases are constitutive of disjunction, etc. According to the inconsistency theorists, what sets the everyday concept of truth apart is that its constitutive principles are inconsistent. Hence, analogously with Dummett's complaints about tonk and classical negation, inconsistency theories identify problems in the very concept of truth. The core idea is that the concept is governed by the naive T-schema or some unrestricted inference rules, such as T-in and T-out.[17] From that assumption, they argue that since the semantic paradoxes lead to contradiction, the involved concept of truth is inconsistent.

Of course, other accounts of the paradoxes can also explain the contradictions, by attributing the problem to a bad *theory* of truth as opposed to a bad concept of truth. It can still be acknowledged that the unrestricted truth principles must be rejected, but it is denied that they have a privileged connection to the very concept of truth. Eklund (2017b) also points out that the semantic paradoxes do not unequivocally point to principles of truth as the culprit, even if these are indeed constitutive. After all, some logical principles, for example of negation or implication, are also involved. Indeed, the naive truth principles on their own do not lead to inconsistency. Classical logic or some suitably strong nonclassical logic is also required, together with a theory of syntax (e.g. arithmetic). So, even if we accept that the unrestricted inference rules are constitutive, the paradoxes need not be a reason for revising the concept of truth. It might equally well suggest a revision of logical concepts.

That brings us to the more general point. Inconsistency theories are just a special case of theories that promote conceptual engineering in logic

---

[16] The most detailed suggestion for a concrete replacement can be found in Scharp (2013).
[17] The T-schema is the biconditional $T(\ulcorner\varphi\urcorner) \leftrightarrow \varphi$, where $\ulcorner\varphi\urcorner$ is the name of the sentence φ. The corresponding proof rules in natural deduction are sometimes called T-in and T-out:

$$\frac{\varphi}{T(\ulcorner\varphi\urcorner)} \qquad \frac{T(\ulcorner\varphi\urcorner)}{\varphi}$$

broadly speaking. The methodology can be extended in two ways, both of which are arguably already implicit in Dummett's work. First, inconsistency need not be the only relevant conceptual flaw that motivates a revision. For Dummett, inconsistency is only a limit case. Less damaging features of the constitutive principles (e.g. nonconservativeness) might also suggest that a logical concept is flawed. Second, conceptual engineering could be applied to logical concepts that are more central to logical theorising than truth.

Consider for example the material implication in classical logic. Although it resembles the English language indicative conditional in a number of respects, it is certainly a minority view that the material implication is an accurate semantic theory of its natural language counterpart. We routinely pause in introductory logic classes to warn our students about straightforwardly translating from natural language 'if' to the the material implication in the formal language. The reason, of course, is that the material implication is at best an idealisation of any natural language concept. We trade in the nuances of the indicative conditional for a number of theoretical advantages. It has a simple semantics and it is interdefinable with other connectives. But most of all the material implication contributes to a highly successful formal theory of mathematical proofs.

In fact, the conceptual engineering methodology need not be limited to connectives and quantifiers. Even the concept of validity, arguably the most central logical concept, can be the target of engineering. In fact, just like inconsistency theorists have been inspired by Tarski's criticism of truth in natural language, we could take Tarski's dismissive remarks about the ordinary concept of consequence as an invitation to think of his model-theoretic definition as a conceptual re-engineering:

> With respect to the clarity of its content the common concept of consequence is in no way superior to other concepts of everyday language. Its extension is not sharply bounded and its usage fluctuates. Any attempt to bring into harmony all possible vague, sometimes contradictory, tendencies which are connected with the use of this concept, is certainly doomed to failure. (Tarski, 1936, p. 409)

Tarski is not trying to account for any concept of consequence that we actually employ in natural languages. Instead he is defining a theoretical concept with the advantage of the precision offered by the model-theoretic framework. Its success is not only due to the fact that Tarski captures important intuitions about logical consequence, but also its utility for mathematical and philosophical purposes.

Eklund (2017c), who supports conceptual engineering in other cases, has suggested it has application in logic as well. According to him, we can best make sense of the philosophical interest in logical pluralism vs monism if we consider the issue as one about which language we ought to have for certain theoretical purposes:

> Doubts about the significance of the actual language project point us toward the normative project. The normative project – what is the best type of language like? what is the best truth predicate like? the best negation? – promises to have the significance that the actual language project lacks. (Eklund 2017c, 5)

What is the upshot of the conceptual engineering approach for the question of disagreement? Given conceptual engineering we can trade in the first-order dispute about logical laws for a metalinguistic (or meta-conceptual) disagreement about whether to adopt, say, intuitionistic negation instead of classical negation. The latter dispute is genuine disagreement, even if it counts as verbal dispute in the sense that it arises wholly in virtue of the parties' dispute about meanings. It is just that the disagreement is not about the actual meaning of a natural language expression, but about which concept we ought to use in our theorising. On this picture, Astrid and Beatrice's dispute about the law of excluded middle cannot be dismissed as insubstantial, but it nonetheless allows for meaning-variance.

6. Theory-choice and concept-choice

The conceptual engineering methodology offers a way to accept the meaning-variance thesis without giving up on genuine disagreement about logic. Yet, we set out to do more than simply find an approach to logical theorising that would vindicate meaning-variance. We also want an approach that combines the meaning-variance thesis with Quine's revisionism and gradualism. Suppose that Dummett's intuitionism is indeed a case of conceptual engineering. Although Dummett and the Quine of 'Two Dogmas' share the view that logic can be revised, their methodologies remain far apart. Dummett's brand of revisionism is a revision of concepts, justified by meaning-theoretic arguments. The resulting justification of basic logical laws is an *a priori* justification. Since Quine rejects apriorism, his revisionism, in contrast, is designed precisely to explain how logical theories can be justified on a nonapriori basis.

The worry, then, is that conceptual engineering is in conflict with Quinean revisionism about logic. However, even if Dummett's revisionism cannot be reconciled with nonapriorism, it does not mean that revision of concepts has no place in a Quinean epistemology of logic. For as we have seen above (in Section 4), Quine thinks that you can have perfectly good reasons for rejecting a logical concept (e.g. classical negation). Those reasons would not be a meaning-theoretic argument like Dummett's, but rather a matter of theory-choice. What is more, Quine's gradualism leads him to conclude that the mechanism of theory-choice in logic ought to be similar to that of the sciences.

Even though Quine (1986) appears dismissive of nonclassical logic, the very chapter where he accuses the deviant logician of 'changing the subject' is followed by a detailed discussion of the advantages and disadvantages of concrete nonclassical theories (Quine 1986, 83–89). His emphasis is on how these theories and classical logic compare wholesale with respect to a number of theoretical virtues (e.g. simplicity, the maxim of minimal mutilation, unification with other theories, expressive power, deductive strength). He readily admits that classical logic has issues both with semantic paradoxes and vagueness, but maintains that the cost of rejecting classical laws in favour of a nonclassical alternative is too high. In Quine (1981) he goes further in listing the drawbacks of a classical theory of vagueness, but concludes that its simplicity still outweighs the advantage a nonclassical logic affords. In sum, he favours an abductive argument for classical logic.

For Quine, there is no conflict between abductive arguments for logic and the meaning-variance thesis. In revising our theories we not only change the laws, but we also introduce new concepts. These concepts should be better suited for the theoretical task, and they are only warranted when they make valuable –or even indispensable – contributions to the theory. Put differently, theory-choice and concept-choice happen simultaneously. In line with his confirmational holism, it is theories and not individual logical laws that receive evidential support. If we reject one theory of logic for another, we not only reject laws of logic, but we replace the theoretical concepts of one logic with another. That is compatible with there being no identifiable set of laws that together form the constitutive principles for a concept in the theory, and thus no identifiable truths of the theory that count as conceptual truths.

Moreover, this revision of concepts is not a feature unique to logical theories. Theory-choice and concept-choice are indistinguishable in general, be it in physics, economics, or logic. Let us return to conceptual engineering for a moment. Recall that many of the examples of allegedly successful applications of the methodology come from outside philosophy (cf. Brun 2016). Eklund (2014, 293) also underlines that there is nothing exceptional about philosophy in this respect:

> [W]hile philosophers often have been concerned with our actual concepts or the properties or relations they stand for, philosophers should also be asking themselves whether these really are the best tools for understanding the relevant aspects of reality, and in many cases consider what preferable replacements might be. Philosophers should be engaged in conceptual engineering. Compare: when physicists study reality they do not hold on to the concepts of folk physics but use concepts better suited to their theoretical purposes. Why should things stand differently with what philosophers study?

Revision of concepts is not a specifically logical or philosophical methodology; it is ubiquitous throughout the sciences. Hence, if meaning-variance is a worry for genuine disagreement in logic, it ought to be one in physics as well. In fact, in her discussion of Quine's meaning-variance thesis in logic, Haack points out that a corresponding meaning-variance thesis has been promoted in the philosophy of science (Feyerabend 1962). Yet, most philosophers would not conclude from this that there is no genuine disagreement between rival theories in chemistry and economics. If the meaning-variance prevents genuine disagreement in logic, that would have to be because the rival parties cannot have rational dispute about concept-choice, for example because of incommensurability. However, the meaning-variance thesis in itself does not provide a reason for such a dramatic conclusion. It also seems unlikely that Quine accepted this conclusion, given his explicit assessment of the advantages and disadvantages of classical and nonclassical logics respectively. But, most importantly, the conclusion is belied by what happens in actual debates about logical theories. The classicist and the nonclassicist do not fail to understand each other's concepts. Williamson and Field understand paracomplete and classical concepts perfectly well, but assess the theories they are embedded in differently.

7. Conclusion

Quine's meaning-variance thesis should not be considered a threat against genuine logical disagreement. The thesis is compatible with revisionism, in fact a form of metalinguistic revisionism already part of many disputes about logic. There is no real conflict between Quine's revisionism and his meaning-variance thesis. Concept-choice is an integrated part of theory-choice, in logic and elsewhere in the sciences.

Acknowledgments

Previous versions of the paper were presented at the 'Engineering Logical Concepts' workshop at the University of Oslo, at the 2nd Buenos Aires-MCMP workshop, LMU Munich, and at the 'Disagreement Within Philosophy' conference at the University of Bonn. I am very grateful to the audiences for many helpful comments.

Disclosure statement

No potential conflict of interest was reported by the author.

Funding

This research is funded by the Research Council of Norway (RCN), grant 251218, and Deutsche Forschungsgemeinschaft (DFG) grant GZ HJ 5/1-1, AOBJ 617612.

References

Arnold, J., and S. Shapiro. 2007. "Where in the (World Wide) Web of Belief Is the Law of Non-contradiction?" *Noûs* 41 (2): 276–297.
Azzouni, J. 2007. "The Inconsistency of Natural Languages: How We Live With It." *Inquiry* 50 (6): 590–605.
Blackburn, S. 1999. *Think: A Compelling Introduction to Philosophy*. Oxford: Oxford University Press.
Brandom, R. B. 1994. *Making it Explicit: Reasoning, Representing and Discursive Commitment*. Cambridge, MA: Harvard University Press.
Brun, G. 2016. "Explication As a Method of Conceptual Re-engineering." *Erkenntnis* 81 (6): 1211–1241.
Burgess, A., and J. P. Burgess. 2011. *Truth*. Princeton: Princeton University Press.
Burgess, A., and D. Plunkett. 2013. "Conceptual Ethics I." *Philosophy Compass* 8 (12): 1091–1101.
Cappelen, H. 2018. *Fixing Language: An Essay on Conceptual Engineering*. Oxford: OUP.
Chalmers, D. J. 2011. "Verbal Disputes." *Philosophical Review* 120 (4): 515–566.
Cook, R. 2002. "Vagueness and Mathematical Precision." *Mind* 111 (442): 225–248.
Davidson, D. 1967. "Truth and Meaning." *Synthese* 17 (1): 304–323.
Dummett, M. 1973. *Frege: Philosophy of Language*. London: Duckworth.
Dummett, M. 1978a. "Is Logic Empirical?" In *Truth and Other Enigmas*, 269–289. Cambridge, MA: Harvard University Press.
Dummett, M. 1978b. *Truth and Other Enigmas*. London: Duckworth.
Dummett, M. 1991. *The Logical Basis of Metaphysics*. Cambridge, MA: Harvard University Press.
Eklund, M. 2002. "Inconsistent Languages." *Philosophy and Phenomenological Research* 64 (2): 251–275.
Eklund, M. 2007. "Meaning-Constitutivity." *Inquiry* 50 (6): 559–574.
Eklund, M. 2014. "Replacing Truth?" In *Metasemantics: New Essays on the Foundations of Meaning*, edited by B. S. Alexis Burgess, 293–310. Oxford: OUP.

Eklund, M. 2017a. *Choosing Normative Concepts*. Oxford: Oxford University Press.
Eklund, M. 2017b. "Inconsistency and Replacement." *Inquiry*: 1–16. Advance online publication. https://doi.org/10.1080/0020174X.2017.1287918.
Eklund, M. 2017c. "Making Sense of Logical Pluralism." *Inquiry*: 1–22. Advance online publication. https://doi.org/10.1080/0020174X.2017.1321499.
Feyerabend, P. K. 1962. "Explanation, Reduction and Empiricism." In *Realism, Rationalism and Scientific Method: Philosophical Papers*, 44–96. Cambridge: Cambridge University Press.
Field, H. 2008. *Saving Truth From Paradox*. Oxford: OUP.
Field, H. 2015. "What is Logical Validity?" In *Foundations of Logical Consequence*, edited by C. Caret and O. T. Hjortland, 33–70. Oxford: OUP.
Haack, S. 1974. *Deviant Logic*. London: Cambridge University Press.
Haack, S. 1978. *Philosophy of Logics*. Cambridge: Cambridge University Press.
Hjortland, O. T. 2012. "Logical Pluralism, Meaning-variance, and Verbal Disputes." *Australasian Journal of Philosophy* 91 (2): 355–373.
Hjortland, O. T. 2014. "Verbal Disputes in Logic: Against Minimalism for Logical Connectives." *Logique et Analyse* 227: 463–486.
Hjortland, O. T. 2017a. "Anti-exceptionalism About Logic." *Philosophical Studies* 174 (3): 631–658.
Hjortland, O. T. 2017b. "Theories of Truth and the Maxim of Minimal Mutilation." *Synthese*. Advance online publication https://doi.org/10.1007/s11229-017-1612-8.
Maddy, P. 2002. "A Naturalistic Look At Logic." *Proceedings and Addresses of the American Philosophical Association* 76 (2): 61–90.
Maddy, P. 2014. *The Logical Must: Wittgenstein on Logic*. New York: Oxford University Press.
McGee, V. 1985. "A Counterexample to Modus Ponens." *Journal of Philosophy* 82 (9): 462–471.
Montague, R. 1973. "The Proper Treatment of Quantification in Ordinary English." In *Formal Semantics – The Essential Readings*, edited by P. Portner and B. Partee, 17–34. Oxford: Blackwell.
Negri, S., and J. von Plato. 2001. *Structural Proof Theory*. Cambridge: Cambridge University Press.
Novaes, C. D., and E. Reck. 2017. "Carnapian Explication, Formalisms As Cognitive Tools, and the Paradox of Adequate Formalization." *Synthese* 194 (1): 195–215.
Paoli, F. 2003. "Quine and Slater on Paraconsistency and Deviance." *Journal of Philosophical Logic* 32 (9): 531–548.
Paoli, F. 2007. "Implicational Paradoxes and the Meaning of Logical Constants." *Australasian Journal of Philosophy* 85 (4): 553–579.
Paoli, F. 2014. "Semantic Minimalism for Logical Constants." *Logique et Analyse* 57 (227): 439–461.
Patterson, D. 2007. "Inconsistency Theories: The Significance of Semantic Ascent." *Inquiry* 50 (6): 575–589.
Patterson, D. 2009. "Inconsistency Theories of Semantic Paradox." *Philosophy and Phenomenological Research* 79 (2): 387–422.
Priest, G. 2006. *Doubt Truth to be a Liar*. Oxford: OUP.

Priest, G. 2014. "Revising Logic." Chap. 12 in *The Metaphysics of Logic*, edited by P. Rush, 211–223. Cambridge: CUP.
Priest, G. 2016. "Logical Disputes and the a Priori." *Logique et Analyse* 59 (236): 347–66.
Prior, A. N. 1960. "The Runabout Inference-ticket." *Analysis* 21 (2): 38–39.
Putnam, H. 1957. "Three-valued Logic." *Philosophical Studies* 8 (5): 73–80.
Putnam, H. 1976. "The Logic of Quantum Mechanics." Chap. 10 in *Mathematics, Matter and Method*, 174–197. Cambridge: Cambridge University Press.
Putnam, H. 1981. "Quantum Mechanics and the Observer." *Erkenntnis* 16 (2): 193–219.
Putnam, H. 2012. "The Curious Story of Quantum Logic." In *Philosophy in the Age of Science: Physics, Mathematics, and Skepticism*, edited by D. M. Mario De Caro, 162–177. Cambridge, MA: Harvard University Press.
Quine, W. V. 1951. "Two Dogmas of Empiricism." *Philosophical Review* 60 (1): 20–43.
Quine, W. V. 1960a. "Carnap and Logical Truth." *Synthese* 12 (4): 350–374.
Quine, W. V. 1960b. *Word and Object*. Cambridge, MA: MIT Press.
Quine, W. 1981. "What Price Bivalence?" *The Journal of Philosophy* 78 (2): 90–95.
Quine, W. V. 1986. *Philosophy of Logic*. 2nd ed. Cambridge, MA: Harvard University Press.
Read, S. 2000. "Harmony and Autonomy in Classical Logic." *Journal of Philosophical Logic* 29 (2): 123–154.
Restall, G. 2002. "Carnap's Tolerance, Language Change, and Logical Pluralism." *Journal of Philosophy* 99 (8): 426–443.
Russell, G. K. 2014. "Metaphysical Analyticity and the Epistemology of Logic." *Philosophical Studies* 171 (1): 161–175.
Russell, G. 2015. "The Justification of the Basic Laws of Logic." *Journal of Philosophical Logic* 44 (6): 793–803.
Russell, G. 2018. "Deviance and Vice: Strength as a Theoretical Virtue in the Epistemology of Logic." *Philosophy and Phenomenological Research*. Advance online publication. https://onlinelibrary.wiley.com/doi/abs/10.1111/phpr.12498.
Scharp, K. 2007. "Replacing Truth." *Inquiry* 50 (6): 606–621.
Scharp, K. 2013. *Replacing Truth*. Oxford: Oxford University Press.
Scharp, K., and S. Shapiro. 2017. "Revising Inconsistent Concepts." In *Reflections on the Liar*, edited by B. Armour-Garb, 257–280. Oxford: Oxford University Press.
Shapiro, S. 1998. "Logical Consequence: Models and Modality." In *Philosophy of Mathematics Today: Proceedings of an International Conference in Munich*, edited by M. Schirn, 131–156. Oxford: Oxford University Press.
Shapiro, S. 2006. *Vagueness in Context*. Oxford: OUP.
Shapiro, S. 2014. *Varieties of Logic*. Oxford: Oxford University Press.
Tanswell, F. S. 2018. "Conceptual Engineering for Mathematical Concepts." *Inquiry* 61 (8): 881–913.
Tarski, A. 1936. "On the Concept of Logical Consequence." In *Logic, Semantics, Metamathematics*, edited by John Corcoran. Indianapolis, IN: Hackett.
Tennant, N. 1987. *Anti-realism and Logic*. Oxford: Oxford University Press.
Tennant, N. 1997. *The Taming of the True*. Oxford: Oxford University Press.
Williamson, T. 2007. *The Philosophy of Philosophy*. Oxford: Blackwell.
Williamson, T. 2017. "Semantic Paradoxes and Abductive Methodology." In *The Relevance of the Liar*, edited by B. Armour-Garb. Oxford: OUP.

Index

Note: Page numbers followed by "n" denote endnotes.

antinomies 5, 6, 102–105, 114, 116, 118–122, 126–132, 146, 147
Aristotelian-Scholastic logic 159
Arnold, J. 198
assertability theory 103, 126, 135
axiomatic second-order logic 10, 22

basic logical laws 199, 201, 202, 211
Beall, J. 175
Bennett, J. 129
Bernays, Paul 94
Bishop, E. 47n2
Blackburn, S. 207
Brandom, R. B. 208n15
Brouwer, L.E.J. 11–13, 41n18, 185
Brun, G. 208n14
Burgess, J. 184, 191

Carnap, R. 180–182
Carnapian resolution 188–191
Chalmers, D. J. 201
classical theories 24, 188, 212
Coffa, J. 182
cognition 107, 121, 148, 149, 159, 166
communication 186, 187
concept-choice 211–213
constructive Mathematics 4, 16, 20, 24, 47, 48, 53, 54
constructive Zermelo Fraenkel set theory (CZF) 5, 24, 47–48, 51–53, 64
constructivism 18, 21, 111, 113, 122
Cook, R. 206n11
cosmological antinomies 118, 120, 121
critical plural logic 5, 68–99
Curry-Howard isomorphism 5, 53, 54, 56–60, 64, 65

Davidson, D. 206
decidability 113, 114
deductive reasoning 176, 180, 182, 188
Descartes, R. 6, 126, 132–134, 136, 147, 161
Dewey, J. 150n12
Dudman, V. 40n17
Dummett, M. 28–43, 29n2, 43, 50, 196, 207, 208

Eklund, M. 209, 211, 213
empirical realism 126, 138, 142, 145
entanglement of logic 3–5, 46, 64, 169
epistemology 133, 142, 148
equinumerosity 119
extensional definiteness 89, 90, 93, 94

false presupposition 114, 116
Field, H. 187, 202
first-order logic 4, 11, 17, 19, 21, 23–25, 50, 70, 88
formal logic 6, 7, 107–110, 156–158, 162–164, 166, 169, 170
formal theory 158, 163, 165, 167, 172, 210
Frege, G. 177–180
Friedman, H. 51n11

generalized union principle 91, 93, 97
general logic 2, 106, 109, 114, 158–161, 165, 166, 171
gradualism 197, 211

Haack, S. 195n1, 196, 199, 203
Hartimo, Mirja 169n11
Heyting, A. 150n13, 185
Hilbert, David 18–21

Hilbert program 16, 17
Husserl, Edmund 7, 157–172

impredicativity 15, 22, 48, 50, 58, 59, 62, 112
infinite pluralities 83, 91
infinity 47, 48, 61, 62, 65, 92, 93, 95–97, 116, 117, 120, 122, 126
intuitionistic logic 4, 24, 28–30, 38, 47, 51, 53, 60, 65, 130–132, 185, 203
intuitionistic type theories 48, 53, 57, 59, 60
intuitionistic Zermelo Fraenkel set theory (IZF) 51

Kant, I. 2, 6, 102, 103, 106, 107, 111, 113–115, 119, 125–149, 157–161, 166, 168, 171
Kantian strategies 136–140
Keefe, R. 175, 176
Kripke, S. A. 127
Kripke models 6, 125–149

Law of Excluded Middle (LEM) 102–105, 107–109, 112, 114, 115, 118, 120, 122, 123
laws of logic 2, 50, 97, 178–180, 182, 188, 197
laws of thought 104, 177–180
Leibniz, G. 126, 132, 134–138, 147, 158, 162, 168
liberalism 83–88
linguistic frameworks 181, 182, 189–191
logical consequence 164, 174, 175, 183, 186, 187, 203, 210
logical expressions 71, 196, 200, 201, 203, 205–207
logical laws 177, 178, 194–201, 203–207
logical pluralism 174–191
logical principles 2, 4, 7, 10, 47, 49, 52, 54, 98, 165, 166, 168, 170, 171, 205
logical terminology 180, 183, 185, 186, 202
logical theories 72, 191, 197, 203, 208, 211–213
Lopez-Escobar, E. G. K. 153n65

MacFarlane, J. 105n6, 174, 178
Maddy, Penelope 101n1
Martin, Gottfried 122n22
Martin-Löf, P. 56, 59, 60, 60n33, 64
Martin-Löf type theory (MLTT) 5, 47–49, 53–59, 61, 63–65
Mathematical antinomies 114, 116, 119, 122

Mathematical statements 4, 38, 57, 102, 110–113, 122
Mathematical truths 20, 39, 42, 146
McGee, V. 201, 204
meaning-variance 198, 203, 205, 206, 211, 213; thesis 194–198, 201, 202, 204–207, 211–213
metalanguage 56, 149
monism 174, 176, 178, 188, 211
Montague, R. 206

natural numbers 4, 12, 13, 22, 23, 32, 34, 54, 55, 57, 64, 68, 91, 92, 120
nonclassical logics 195, 196, 200, 201, 206, 212, 213
normativity 174–191; of logic 7, 174, 180, 183, 188, 190, 191

ontological/ontology 133–135, 138, 142, 148, 149; intuitionism 4, 31, 32, 37, 39, 43; route 4, 28–30, 32, 36–38, 43

Paoli, F. 203
Parsons, Charles 2, 102n2, 110, 168
plural comprehension scheme 88, 89
pluralism 175, 176, 180, 187, 190
pluralities 5, 71, 76–81, 83–85, 88–97; proper classes as 81–83
plural logic 5, 68, 69, 71, 72, 74, 76, 82, 83, 95–98; vs a simple set theory 72–75
Poincaré, H. 48, 61, 63
Poincaré's logic of infinity 61–63
Posy, C. J. 103, 149n1, 153n58
powerplurality 94, 96, 97; principle 94, 96
predicative law of excluding the middle (PLEM) 104, 105, 108, 112, 114
predicativity 5, 48, 57, 59–61, 65
Priest, G. 200, 205
Principle of Complete Determination (PCD) 104, 108–110, 114, 118, 121, 122
pristine epistemology 134
pristine ontology 134
pristine semantics 135
propositions 6, 19, 20, 31, 47, 48, 54–60, 165, 174, 196
psychologism 30, 107
Putnam, H. 150n17, 202, 203

quasi-combinatorial reasoning 94, 97
Quine, W. V. 195, 196, 205, 212

Read, S. 175, 176
reasoning 3, 4, 10, 23–25, 47, 96, 97, 116, 164–166, 168, 180

replacement 82, 92, 93, 95, 213
Restall, G. 175, 203
revisionism 197, 198, 202, 205–207, 211, 213

Scharp, K. 209n16
Searle, J. 179
second-order logic 3, 22–24
sensory perceptions 136–138
sets 54–58, 75–79; theory 18, 22–25, 46, 48, 50, 64, 65, 80, 82, 96, 97, 121
Shapiro, S. 198, 202, 206, 207
Sher, Gila 160, 171n12
singleton pluralities 90
Sommers, Fred 104
special logic 106, 107, 148
Steinberger, F. 177, 177n5
subpluralities 93, 94, 96
superpluralities 93, 94, 96

Tennant, N. 185, 208n15
theory-choice 211–213
traditional metaphysics 2, 7, 114, 189, 191
traditional plural logic 5, 69, 71, 72, 80, 88, 93
transcendental criticism 158
transcendental idealism 145–146
transcendental logic 106, 107, 109, 110, 114, 119, 156–158, 169, 171
transcendental philosophy 112–114, 142, 143, 146, 149, 157, 160, 167–169
transcendental problems 160, 168
transcendental realism 114, 129, 130, 146
transcendental scrutiny 157, 165, 168, 169
truth values 31, 32, 38, 39, 59, 60, 75

universal plurality 5, 88, 89, 93

verbal disputes 195, 201, 207, 211

Weyl, Hermann 13–18, 21, 48n6
Williamson, Timothy 5, 198n2, 204
Wolff, Michael 106
Wright, C. 42n20

Zermelo Fraenkel set theory (ZF) 46–52, 62, 64

9781032573540